Praise for *UnSavory*

"A fascinating memoir of a man whom history may well regard as one of the most influential of his generation. This is a must read for anyone who cares about our environment and its ability to sustain us."
—Jamie Blackett, author of the monthly column, "Farming Life" for *Country Life* (U.K.)

"A man of modesty to the point of self-deprecation, Allan does not hesitate to thrash through the thorniest problem and break from convention to find a solution. He will remind you that we all function within the ecological principles that apply to every plant and animal. *UnSavory* is Allan Savory in his own words, a breathtaking compendium of stories that exalt, move, and intrigue—some funny, some exhilarating, some tragic."
—Gretel Ehrlich, author of *The Solace of Open Spaces*

"Savory freely admits his life has been a mistake-ridden journey—as is generally the case with outstanding contributors. Yet those mistakes have led to the regeneration of vast swathes of denuded landscapes across the world. In time, Allan Savory—ecologist, warrior, and visionary—will be properly recognized as one of The Greats."
—Dr. Charles Massy, farmer, scientist, and author of *Call of the Reed Warbler: A New Agriculture, A New Earth*

"An extraordinary tale from a truly extraordinary human being. I daresay, you may discover and learn more about the Earth (and the earth) than you could ever have imagined."
—Paul Hawken, author of *Carbon: The Book of Life*

"Thousands of us around the world attribute much of our soil-building and profitability success to Holistic Management. Here is the prequel to all that discovery. Allan lets us "track" with him through his formative early life—where character, conviction, commitment, and consequences converge to make the man so many of us admire."
—Joel Salatin, farmer, Food Freedom advocate, and author of *Everything I Want to Do Is Illegal*

"I love the title *UnSavory,* as Allan has been condemned by academics, much as Galileo was labelled a heretic for proving Earth rotates around the Sun, and the Wright brothers initially ignored because academics insisted powered flight was impossible. As a result of Allan's influence, millions of acres worldwide have been regenerated. History will favor Allan, and his proven, practice-based management will be regarded with the same significance as Galileo's and the Wright brothers' achievements."
—Dr. André Leu, international director, Regeneration International

"Savory was, and is, ahead of his time politically, militarily, and environmentally. In *UnSavory,* the life and insights he describes are powerful, heartbreaking, thought-provoking, and honest."
—John Padbury, former Rhodesian Special Branch detective inspector and author of *Battle for Hurungwe*

"*UnSavory* offers a rare insight into the remarkable life of one of mankind's great adventurers. There is much in this book to be learned. There is even more to be admired."
—Will Harris, owner and (4th generation) manager, White Oak Pastures

"Since my childhood days in Zimbabwe I have heard the name Allan Savory, always in the same context: the place where conflict on and about the land has intersected with uncompromising truth. Through the decades, the nature of this place has varied, whether it be military, political, or environmental. But the constant has been this man's courage—spoken, written, and lived—to bring into being his hard-fought solutions for the loss of human and wildlife habitat. What an inspiration *UnSavory* is for those who follow."
—Larry Norton, internationally recognized wildlife artist

"*UnSavory* is filled with terrifying and joyous encounters with elephants, lions, leopards, hippos, rhinos, crocodiles, buffaloes, and mambas, and cobras. There are also human moments of love, generosity, greed, and stupidity. More than most, Allan has worked to craft actual solutions to our very real dilemmas. And now in this book he tries to pass on that learned wisdom, like sharing a flickering torch that might light our way. He preaches a revolution of the heart."
—Dave Chapman, co-director, Real Organic Project

"May we all be inspired by this extraordinary journey of discovery; not a discovery of a new planet or mathematical truth, but a discovery of a new way to *see*. Savory's life is one of genuine meaning and experiences woven together by the all-too-rare thread of integrity, courage, and grit—softened by a mischievous wit."
—John Fullerton, president, Capital Institute

"Historical figures who have truly original insights typically have enthralling back stories, and Allan Savory is no exception. Service comes from the hearts of too few and leadership in service means holding true no matter what. That is the lesson for me of *UnSavory*."
—Darren Doherty, founder, Regrarians (Australia)

"*UnSavory* is a gift to humanity in our hour of need: an invaluable, urgent guide in the face of climate collapse. Set against the political flux at the end of colonial southern Africa, it narrates how Allan Savory came to formulate his revolutionary Holistic Management through sustained, attentive observation of the veld—first as a game ranger, then as a soldier in a bush war, and as a farmer—and recounts the daunting obstacles he faced. Read this book if you, like me, need affirmation that we are allowed to hope—if we act now."
—Clifford Bestall, international filmmaker; director of the award-winning documentary, *The 16th Man*

If you can bear to hear the truth you've spoken
Twisted by knaves to make a trap for fools,
Or watch the things you gave your life to, broken,
And stoop and build 'em up with worn-out tools . . .
–Rudyard Kipling

Out of intense complexities, intense simplicities emerge.
–Winston S. Churchill

UNSAVORY

UNSAVORY

African Stories of Wildlife, War, and the Birth of Holistic Management

ALLAN SAVORY

Foreword by Gretel Ehrlich

SAVORY INSTITUTE

Boulder/Colorado/United States

Savory Institute
c/o Permissions
855 Arapahoe Avenue
Boulder, Colorado 80302 USA
contact@savory.global

Library of Congress Control Number: 2026903479

ISBN-13: 979-8-9935454-0-0
eISBN: 979-8-9935454-1-7

Editor: Jamie Ann Cross
Cover Design and Maps: Sarah Selkirk Dodge
Interior Design and Composition: Wanda España, Wee Design Group
Author Photographer (Back Cover Photo): Igor Potgieter

Printed in the United States of America
25 26 27 28 29 1 2 3 4 5

For my mentors—wise, courageous, and not without fault,
and for Jody

BRIEF CONTENTS

CONTENTS

FOREWORD

Allan Savory grew up in the African bush. He lived simply then and now. He is both a deep thinker and a doer, and his belief in common sense and on-the-spot observation is so acute as to be uncannily prescient. Quiet and self-effacing, with an avidly curious mind, Allan is a man of amazing latitude, empathy, and both physical and moral courage. He is also mischievous, deeply disciplined, and ridiculously tenderhearted. He abhors the least hint of racism and is driven to tears by mistreatment of animals. He has a sardonic wit and is a practical joker.

In 1987, when *Time Magazine* asked me to write essays about "visionary westerners," I thought immediately of Allan Savory. He was not, of course, a "westerner" in the American sense, but his influence on the American West had been profound. His discovery that livestock were essential to reversing desertification and regenerating the world's grasslands was—and still is—controversial, and in those early days western ranchers and environmentalists talked of little else. To write the magazine assignment, I took Allan's weeklong course and had the first of many private interviews with him. I had my own small cattle ranch in northern Wyoming and had sensed something important was missing. I hadn't truly grasped what it was until taking Allan's life-changing course. Working with land and livestock now required me to be a devoted student of the land. It demanded intimacy, flexibility, and a keen awareness. I learned to manage for abundance; I learned to listen to the truth the land tells you.

As this book attests, Allan has affected the lives of thousands of ranchers, farmers, villagers, pastoralists, environmentalists, government workers, kings, and queens. He has challenged the thinking of academics and made a lot of people mad. But the teachers—who learned what some perceive as radical ideas—are teaching new teachers. "It's not radical, it's just common sense," he'll tell you. "To teach people how to see, how to think holistically, how to decide, how to act in accordance with the laws of the natural world of which we are a part, is profoundly important," he says. To that end Allan and his supporters have created the Savory Institute and in turn a network of regional learning centers where holistic management (as his work has become known) is taught and local communities of practitioners are formed.

This book is Allan Savory in his own words, a breathtaking compendium of stories that exalt, move, and intrigue—some funny, some exhilarating, some tragic. Ecologist, philosopher, teacher, warrior, tracker, parliamentarian, and friend, he aims at nothing less than keeping us from the abyss, nothing less than reversing the loss of biodiversity and the desertification that is symptomatic of it, so that Earth continues to remain habitable by humans. And if you are a friend in distress, he will help you.

What has always distinguished Allan is his early willingness to declare utter ignorance when faced with new problems and his determination to discover their root causes. He has never been too proud to say he doesn't know or is wrong. Instead, he adopts a "beginner's mind," steps back far enough to see the whole context, and observes how the patterns of nature work within it. Outraged when humans resist exploring new ideas, he is never afraid to be controversial or to break conventional boundaries.

Sitting with Allan in his bush camp in Zimbabwe, I see him observing the life around him. It is early morning, and the francolins greet us with their ear-piercing cackle. The night before, a lion roared and salmon-colored sand from the Kalahari dusted our chairs. As the sun rises, a giraffe wanders by and two impala drink at the river. Baboons make a racket, a warthog climbs up from its hole, a guinea fowl scurries between trees. Sipping tea, Allan's quietude gives way to a sudden profundity, joke, or a story he forgot to tell the night before. A man of modesty to the point of self-deprecation, he does not hesitate to thrash through the thorniest problem and break from convention to find a solution. He will remind you, over and over, that we all function within the ecological principles that apply to every plant and animal. Barefoot, he sits in stillness, imbued with what might be called an African equanimity. Allan Savory is a man with deep empathy, an unassailable moral compass, and a hunger for justice, who never loses his sense of humor or his cool.

—Gretel Ehrlich, author of *The Solace of Open Spaces*

PREFACE

People dream of colonizing space or mining the moon. Trillions of dollars are spent on these dreams as well as on entertainment, the international arms race, and wars. Yet, we spend very little to address the greatest problem facing humanity: the accelerating destruction of Earth, the only habitable planet known to us.

This memoir is the story of the first fifty years of my life. It describes how it happened that I, as a young ecologist planning to devote my time on Earth to saving the wildlife I loved, turned instead to solving the riddle of their habitat loss (a far greater threat than our focus on poachers). By the end of my story, you may better understand why we were so blind to the damage right before our eyes and why this same blindness plagues us to this day, some seventy years later, as evidenced in the dramatic loss of biodiversity, the desertification that is symptomatic of it, and the role both play in spiraling climate changes.

Nevertheless, there is hope in my story. It's unrealistic to think we will find a silver-bullet solution that will address climate change, even if we all eat vegan, stop burning fossil fuels, and cook our food in solar ovens. There is, however, a way to address the problem at its root cause—and this is the story of how that has become possible.

My story unfolds amidst an odyssey of misunderstandings, mistakes, institutional opposition, false promises, racism, violence, and war. Like others who challenge beliefs and dogma, I've been branded by academics with a variety of unsavory terms and proven "wrong" in countless peer-reviewed papers and books, though one paper should have been sufficient to prove me wrong.[1] This is indeed frustrating, but I've since learned to embrace the condemnation as yet more proof that the insights I share here are truly revolutionary and understandably infuriating to established sages.

In 1983 the efforts and observations of many people gave birth to Holistic Management, as my work has become known. Holistic Management is the solution for saving ourselves and our habitat, including our climate. It is the fruit of a growing group consciousness and could never have been developed by any one person, in any one country, or by any single university, government, corporation, environmental organization, or in any one branch of modern science. Nor, I believe, could it have been developed at any time in history other than during the unique situation that arose in Africa in the waning days of the British Empire.

The language and the place names that figure largely in this story have changed over the years. I could have sanitized some of that language as well as used modern place

[1] I'm paraphrasing a quote attributed to Albert Einstein when he learned a book authored by 100 scientists had proved his theory of relativity "wrong." Einstein famously replied, "Why 100? If I was wrong, one would be enough."

names but chose instead to use the words we used at the time (I've included a list of old and new place names at the beginning of this book). Of note is my reference to people as Africans, Afrikaners, Europeans, or Coloreds.

When I use the term "Africans," I am referring to the descendants of Bantu[2] ancestors who originated in West Africa centuries before, migrating first across and then down Africa. These "Africans" displaced the earliest San people, often called Bushmen, who left their art and stone artifacts all over Rhodesia. Distinct from these Black Africans are the White "Afrikaners," descendants of the Dutch settlers who arrived in South Africa's Cape region in 1652, some of whom later migrated north into Rhodesia. Although my family's roots go back four generations in Africa and I have never seen myself as anything but African, I am white. Whites generally referred to themselves as Europeans (Europe being our most recent origin). People of mixed African and European or Afrikaner blood were referred to as Colored and treated as White legally but not socially.

In 1923, Southern Rhodesia became a country and self-governing colony. I was born a mere twelve years later in 1935. Southern Rhodesia had evolved into a modern nation state—albeit one of colonial character—that included a British-style government, modern infrastructure, and a robust economy. Little did I know at that time that my beloved country would soon spiral into a turbulent, violent nation—renamed Rhodesia in 1964—under the racist regime of Prime Minister Ian Smith.

In 1980, Rhodesia became Zimbabwe, and we all became Zimbabweans, regardless of race or tribe. And that is how I refer to my fellow countrymen and women, and how I see myself today. I use the terms "Black," "White" and "Colored" throughout because I cannot tell the story of racism as I experienced it in Southern Africa without using them.

My corner of Africa has rapidly transformed, but alas so has the world. Recognition of environmental and social issues has opened public dialogue in ways unimaginable in the mid-1900s. Unfortunately, many of today's historians justifiably accuse my generation and its antecedents of now unforgiveable sins. Present understandings can also, however, misconstrue past events and blind us to the context that gave past events their meaning. I read history for what it can tell us about our current condition, but acknowledging its seeds in fact were planted "back then" but often unrecognized.

Many readers may choke on the pages I devote to hunting and to guns, considering this yet another "unsavory" aspect of my life. In the wild Africa of my youth, a rifle was essential for the kind of life I led. Therein lies much of my tale, and even now when I hardly have breath to climb my stairs to bed, just a whiff of Young's 303—the solvent we used to clean rifles at the end of a day in the bush—will reignite my love for wild-

---

[2] Bantu were named for the Bantu family of indigenous languages they spoke (languages that dominate much of sub-Saharan Africa today). The South African apartheid regime appropriated this term in creating "Bantustans"—places where Black South Africans were forcibly resettled away from "White" areas.

life and the extraordinary sights, sounds, and rhythms that first drew me into the game department. My relationship with the animals I hunted was a spiritual one not unlike that of hunters hundreds of thousands of years before me. It's a relationship rarely seen today outside the remnant traditions of indigenous North Americans, Mongolian eagle hunters, and Aborigines.

In the context of my place and time, becoming a hunter was to enter the society of Nature as a member of the noble peerage of predators—lions, leopards, crocodiles, and falcons—and it carried certain rights *and* responsibilities. Today, people with little connection to that world tend to demonize even the most ethical hunters like me, preferring that all animals die naturally. I've witnessed my share of natural deaths. While I strived for solutions to habitat loss, thousands of animals died of starvation, disease, accidents or, if lucky, in the jaws of a crocodile or lion—the most humane death nature offers.

While writing this memoir decades after the events described, I found on rereading my diaries that some of my clearest memories were faulty. This led to many months of fact-checking aided by letters, flight logs, parliamentary records, and boxes of newspaper cuttings my family had kept over the years.

I also found I couldn't tell my story in strict chronology because of my overlap in roles. After my early years as a game ranger and biologist, I was simultaneously a farmer, game rancher, independent scientist, ecological consultant, army officer, and political leader. For clarity's sake, I organized the chapters into sections based on these roles. This means there is some jumping around in dates, particularly in later chapters, which can be confusing—a problem I hope I've resolved with the included timeline.

I also realized that I could never have written this memoir until now; it has taken me almost nine decades to understand my life well enough to clearly tell of my journey. Detailing my progress through life has challenged me to make sense out of small signs in much the same way as tracking a wild animal through the bush. I've learned from native masters, both animal and human, that "seeing" a path demands much more than simply noticing tracks in the mud. That said, this is a tale on many levels: tracking a tracker who tracked a mystery—the mystery of how and why my own kind fails to "see" that we are following a path to destruction when we could take another path to avoid it.

A. SOUTHERN AFRICA
(Showing the Central African Federation)

B. SOUTH AFRICA

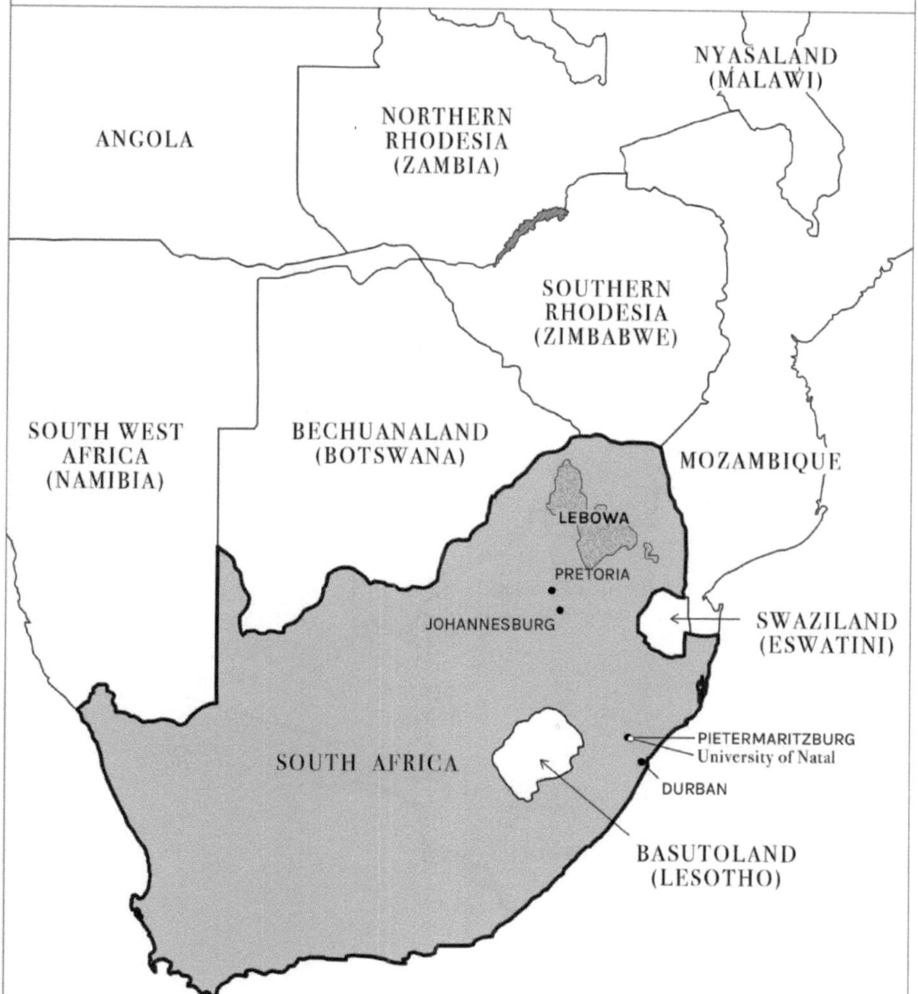

C. NORTHERN RHODESIA
(Showing the Six Provinces)

D. NORTHERN PROVINCE
(One of the Six Provinces of Northern Rhodesia)

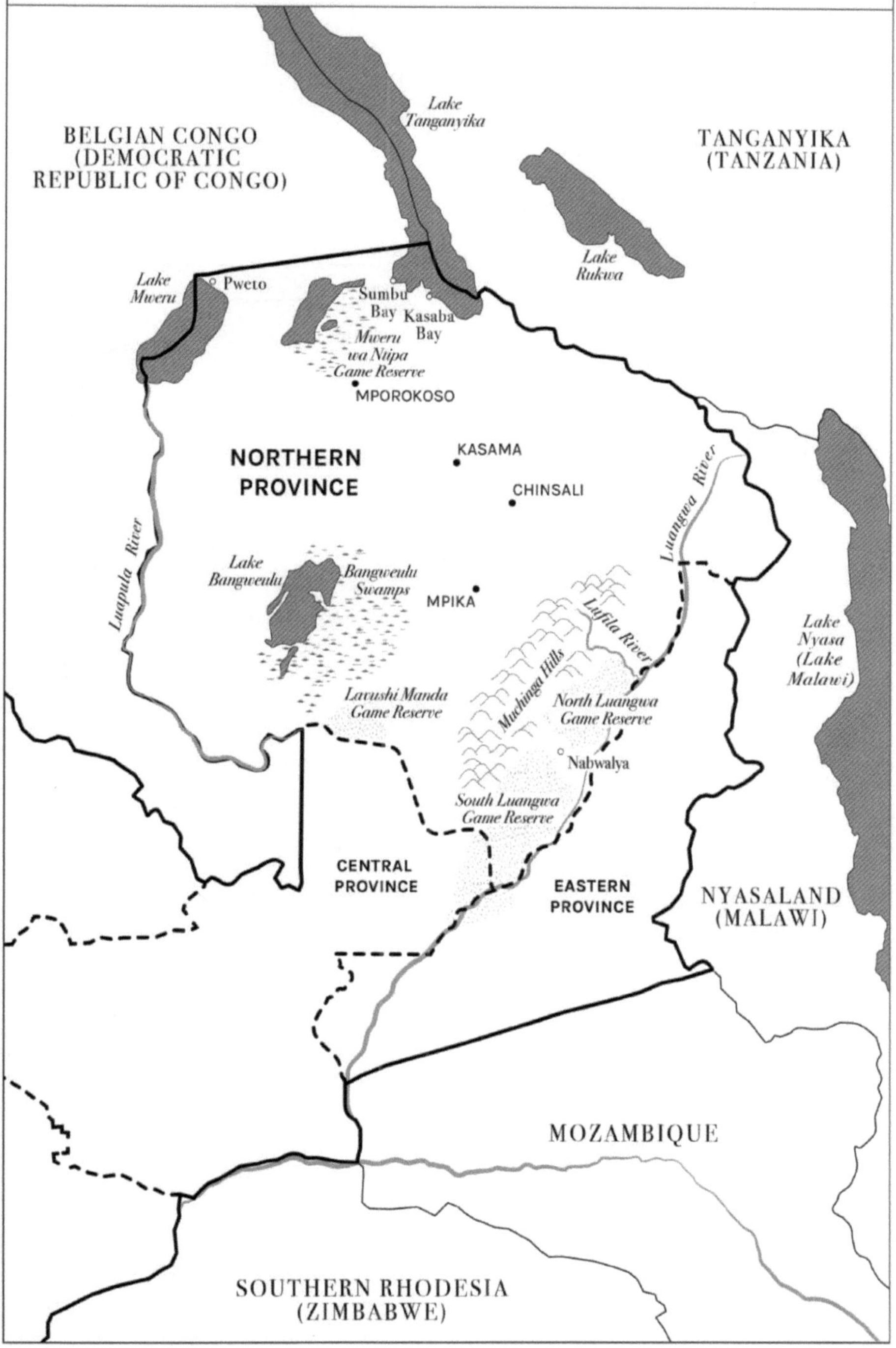

E. SOUTHERN RHODESIA

TIMELINE

A list of both historical events in Southern Rhodesia (now Zimbabwe) and Allan Savory's personal milestones addressed in this book.

1923: Southern Rhodesia becomes a country and self-governing colony.

1933–1953: Godfrey Huggins, 1st Viscount Malvern becomes the fourth and longest-serving prime minister of Southern Rhodesia (and the longest-serving prime minister in British Commonwealth history).

1935: Allan is born in Bulawayo, Southern Rhodesia.

1943–1947: Allan attends junior school, mostly Milton Junior.

1948–1952: Allan attends senior school at Plumtree.

1953–1955: Allan attends University of Natal in Pietermaritzburg, South Africa, earning a Bachelor of Science in Zoology and Botany.

1953: The Federation—Central African Federation—is formed in August (consisting of Northern Rhodesia, Southern Rhodesia, and Nyasaland).

1953–1956: Huggins steps down as prime minister of Southern Rhodesia to serve as the first prime minister of the Federation.

1953–1958: Garfield Todd serves as Southern Rhodesia's prime minister.

1956–1958: Allan works for the Northern Rhodesian Game Department, first as game ranger then provincial game officer.

1956–1963: Sir Roland "Roy" Welensky serves as the second and last prime minister of the Federation.

1957: Allan and Shirley Garbutt marry.

1958–1962: Edgar Whitehead serves as Southern Rhodesia's prime minister.

1958: Allan's first son Jim is born on April 1 in Kasama, Northern Rhodesia.

1958–1964: The wildlife rescue operation, Operation Noah, takes place on the Zambezi River.

1959: While waiting for a game department post, Allan works for the tsetse department in Southern Rhodesia as an entomologist.

1959: The Nkhata Bay Incident occurs on a pier in Lake Nyasa, Nyasaland on March 3. This is the first use of force in the region since Sir Harry Johnston defeated the slave-trading Yao Chiefs in 1895.

1959: The Federal Prime Minister declares a "Nyasaland Emergency" and calls up 1st Battalion Royal Rhodesia Regiment to support the police in Operation Sunrise (March 3–5), which results in mass arrests of Black activist leaders.

1960–1963: Allan works for the Southern Rhodesian Game Department, first as quelea control officer and then as head of Wildlife Research and Management.

1960: British Prime Minister Harold McMillan delivers his "Wind of Change" speech to the South African Parliament in Cape Town, South Africa on February 3; this marks the beginning of the end for the Federation.

1960: Allan's first daughter Megan is born on May 26 in Salisbury, capital of Southern Rhodesia.

1961: Southern Rhodesia's 1961 Constitution is approved in July.

1961: The nontribal political party Zimbabwe African People's Union (ZAPU) is founded in December, campaigning for Black majority rule in Rhodesia.

1962: Allan collaborates behind the scenes in forming the Game Ranchers' Association in Southern Rhodesia.

1962: In December, a new conservative political party, the Rhodesian Front (RF), led by Winston Field, wins the majority of seats in Parliament following a blatantly racist election campaign.

1962–1964: Winston Field's short-lived term as prime minister of Southern Rhodesia.

1963–1970: The duration of the game ranching company, Wildlife Utilization Services Pvt. Ltd. (WUS).

1963: Allan buys Kapombe Farm (sells it in 1965).

1963: In August, the Zimbabwe African National Union (ZANU) is founded—a tribal (Shona) party that opposes the White minority rule of Rhodesia.

1963: Britain dissolves the Federation on December 31.

1964: Nyasaland is renamed Malawi, Northern Rhodesia is renamed Zambia, and Southern Rhodesia (no longer having to distinguish itself from Northern) is renamed Rhodesia. Both Malawi and Zambia become independent republics. Rhodesia remains a self-governing colony.

1964–1979: Ian Smith serves as Rhodesia's prime minister, replacing Winston Field in an internal coup.

1964–1979: Civil conflict escalates into the Rhodesian Bush War (ultimately leading to the end of White minority rule in 1980).

1965: Under the Unilateral Declaration of Independence (UDI), Rhodesia declares independence from Britain in November.

1965–1966: Allan trains Guerrilla Anti-Terrorist Units (GATU) in guerrilla warfare.

1966: United Nations sanctions on Southern Rhodesia begin.

1966: Evelyn and Edward Rushmore, Allan's first clients in his consulting business, visit Allan's home in Bulawayo to ask for his help with their ranch.

1967: Allan buys Kazuma and makes it his family's home.

1967: Allan forms the Tracker Combat Unit (TCU).

1968: Allan is elected a member of Parliament (MP, Matobo constituency) in Smith's Rhodesian Front Party (RF).

1968: Allan's second son Hugh is born in March in Bulawayo, Rhodesia, and dies in September.

1968: Allan serves as a witness for Tsumis Ranch, Namibia, helping the owner Malcolm Kroon win what became known as the Tsumis Case.

1969–1975: The duration of the Charter Estate Grazing Trial.

1969: A referendum is held in June; Rhodesia's new constitution is approved (replacing the 1961 Constitution), as well as a republic is established, paving the way for White rule.

1970: The first "Advanced Project" begins at Liebig's Ranch (the second Advanced Project begins a year later at Igava Ranch).

1970: Allan's third son Rodger is born on May 16 in Bulawayo, Rhodesia.

1972: Allan crosses the floor in Parliament, forming the Rhodesia Party (RP) and leading it until 1974.

1972: The Southern Rhodesian government expropriates Kazuma.

1972: As MP, Allan keeps the "Matetsi Debate" (a private members' debate) going for two weeks—from July into August—forcing the government to form Kazuma Pan National Park, which they did in 1975.

1973: In January, Prime Minister Ian Smith closes Southern Rhodesia's border with Zambia.

1973: TCU is absorbed into a new regiment, the Selous Scouts.

1973: Allan's second daughter Claire is born on November 24 in Salisbury, Rhodesia.

1974: Allan and Shirley divorce.

1974: The Rhodesia Party (RP) runs a candidate against Allan; in July the vote is split, and Allan loses his seat, leaving the Rhodesian Front Party (RF) with all fifty seats.

1975: Allan and Janet Barnes marry.

1976: Chief Jeremiah Chirau forms the political party, ZUPO (Zimbabwe United People's Organization)—a party ultimately controlled by Prime Minister Smith and disbanded in 1980.

1976: Allan's third daughter Sarah is born on October 1 in Salisbury, Rhodesia.

1976: In October, Allan joins African nationalist leaders in Geneva, Switzerland at the beginning of the Geneva Conference for attempted peace talks.

1977: Allan forms and then leads the National Unifying Force (NUF), an amalgam of all the White opposition parties in Rhodesia.

1977: In Rhodesia's August general election, no candidates fielded by the NUF win a seat; Allan boycotts the election.

1978: Allan speaks in January at the International Stockman's School in Phoenix, Arizona (and gives another talk in the United States the following year in San Antonio, Texas).

1978: The Internal Settlement is signed on March 3, 1978, Smith's government promising general elections in 1979 for a transitional government.

1978: In October, Allan goes to Washington, D.C. on behalf of the NUF to inform the United States of the "other side" of Prime Minister Smith's Internal Settlement.

1979: In February, Durban's newspaper, the *Sunday Tribune,* publishes two articles on Allan: "The Guerrilla War Smithy Just Can't Win" and "Search Warrant After Interview with Anti-Smith Man. Police Seize Savory Tapes from Tribune."

1979: On March 30, Allan escapes Rhodesia, going into exile in the Cayman Islands.

1979: On May 30, Bishop Muzorewa is sworn in as Rhodesia's prime minister; Smith named to serve as minister without portfolio. They take office in June. This only lasts until December of the same year.

1979: Rhodesia's name is again changed, this time to Zimbabwe-Rhodesia.

1979: The Lancaster House Agreement is signed in December, effectively concluding the Rhodesian Bush War.

1980: Robert Mugabe's party, ZANU-PF, wins general elections by a landslide in February.

1980: Zimbabwe-Rhodesia becomes Zimbabwe in April; Mugabe becomes Zimbabwe's prime minister.

1981: Allan and Janet divorce.

1982: Allan and Jody Butterfield marry.

1983: Allan, with the help of those attending his courses, develops his ideas into a management framework that facilitates the practice of what he names *Holistic Management.*

1984: Allan and Jody co-found the nonprofit Center for Holistic Resource Management and in 2009, with six others, co-found the Savory Institute.

OLD AND NEW PLACE NAMES

Some names in this book have changed throughout the years; for quick reference, below is a list of the most prominent ones. The new name is included in the text at first mention.

Old Name	**New Name**
Bechuanaland	Botswana
Belgian Congo	Democratic Republic of the Congo
Chipinga	Chipinge
Gatooma	Kadoma
Gwelo	Gweru
Hartley	Chegutu
Lake Nyasa	Lake Malawi
Lundi River	Runde River
Marandellas	Marondera
Northern Rhodesia	Zambia
Nyasaland	Malawi
Que Que	Kwekwe
Sabi River	Save River
Salisbury	Harare
Southern Rhodesia	Zimbabwe (Rhodesia 1964–1979)
South West Africa	Namibia
Swaziland	Eswatini
Tanganyika	Tanzania
Umtali	Mutare
Wankie National Park	Hwange National Park

PART I SCHOOL DAYS

CHAPTER 1 THE SMALL BOY

Childhood Lessons in Reliance, Responsibility, and Being an Underdog

1935–1948

CHAPTER 2 THE "GAME" OF LEARNING

Adolescent Lessons in Building Character and Breaking Rules

1948–1952

CHAPTER 3 THE "WORK" OF LEARNING

University Lessons in Love, Mambas, and Mud

1953–1955

CHAPTER 1

THE SMALL BOY

Childhood Lessons in Reliance, Responsibility, and Being an Underdog

1935–1948

And so I became a lifelong defender of underdogs.

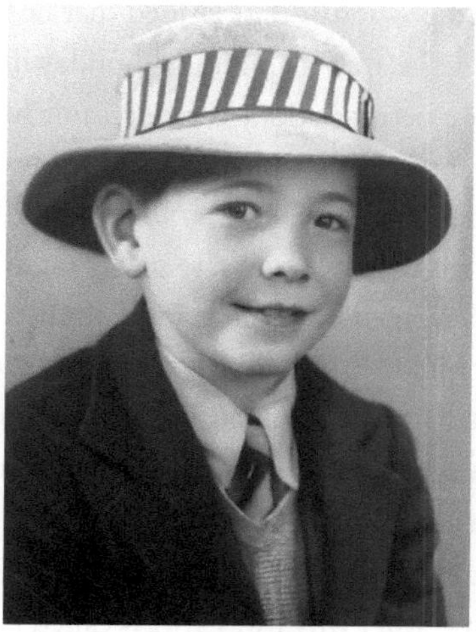

(*left*) My school photo while attending Milton Junior (1943). (*right*) I'm fishing on the upper Zambezi River (1946).

My first clear memory in life is at age three, lying on my back in the corner of a hospital ward in Bulawayo, Southern Rhodesia (today Zimbabwe) with my right leg in a plaster cast and a football bladder suspended from the ceiling so that I could exercise by hitting it with my hands. The long days were punctuated by visits from my mother, Elaine, and my sister, Ann, who was five years older. I would listen to the stories they read to me. That was my world as I became conscious of it. I was one of the many children infected by polio in those days before that terrible disease was almost wiped out. After eighteen months in plaster, I was fitted with an iron leg brace and could begin to stagger around on my own, considering myself lucky to only be suffering from one partially paralyzed limb.

My great-grandfather, William Henry Savory, had arrived in Durban, South Africa, from Norfolk, England, on the SS *Aliwal* in September 1849 and made it his home. My grandfather, Clifford Redin Savory, was born a decade later and moved between Durban and Johannesburg working as an administrator in various businesses. He married and fathered four sons and two daughters before being widowed shortly after my father's birth.

My grandfather felt it was his responsibility to tell his sons which professions they would pursue. The two elder brothers, when they returned from France at the end of World War I, did as they were told. Bryan become a lawyer and Cecil an accountant. The two younger brothers, Guy and my father James, or Jimmy, would also be expected to acquiesce to their father's wishes. Guy, however, had previously determined that he would decide his own fate. He'd gone to Northern Rhodesia (now Zambia) on holiday in 1919 and refused to return, seeing life as a farmer there a better option. When it came his turn to discuss his future career, my father, Jimmy, wanted to be a horticulturalist but could not think of the word and did not want to say "gardener," so he was told he would be an engineer. My father became the first engineering graduate from the newly established University of Natal[1] in 1925 and was soon after engaged by Pauling and Company on a survey for the Rhodesia railways, which brought him to Southern Rhodesia. My mother, Elaine Verna Buchanan Muir (her family originally from Scotland), also came up from Natal shortly after as a teacher and two years later married my father on Guy's farm in Ndola (Northern Rhodesia). My sister was born in 1930 and I in 1935.

Though my parents deemed my full name to be Clifford Allan Redin Savory, I was called Allan from day one. Later, when I asked my father why I was always called Allan, I found his reasoning to be quite humorous. He said my mother and he wanted to call me Allan Redin Savory, but, could not possibly send me to boarding school—where boys always gave each other nicknames—with the initials ARS (my nickname would almost certainly then be *Arse*!). So, they added Clifford. When I then asked why they didn't call me Clifford, he replied, "Well, you would've gone to school with the name Clifford—and some boy would have abbreviated that to Cliff—and then one would have used Precipice and that would have been changed to *Piss*!"

My father was away much of the time in my early years because his work as a civil engineer entailed being in the bush surveying the railway lines, roads, and dams of our young country. When I was just four years old and still recovering from polio, Britain declared war on Germany and my father left to join the Royal Engineers in North Africa and the Middle East. My father always maintained that while my mother and he were deeply concerned with me starting life as a cripple (the effects of polio leaving my right leg shorter than my left), he left for the war feeling I was going to be all right. What made him so confident, he

[1] The University of Natal in South Africa was named for the province in which it was based. In 2004, it was renamed the University of Kwazulu-Natal.

said, was what he saw out the window one day as I played with a group of boys on a sand heap. As I struggled to the top, dragging my iron-clad leg behind me, then slid down belly first with the rest of the boys, I was not behaving at all like a cripple.

From day one, World War II dominated our lives. Southern Rhodesia was the first country to declare war against Germany in support of Britain, even before Australia and Canada. I have a vague memory of a group of war-bound men, including my father and my uncle Guy, in their Rhodesian bush hats pulling out of Bulawayo station. Later I have memories of watching endless flights of Spitfires and large bombers as many of the Royal Air Force pilots trained in the skies of distant and fair-weather Southern Rhodesia.

My mother moved us to Southern Rhodesia's capital city, Salisbury, where we stayed in a boarding house at 14 North Avenue. Mrs. Stubbs, who ran it, treated me like one of her own while my mother joined the army and became a sergeant at one of two prisoner-of-war camps outside of Salisbury. The two camps, known as Tanganyika One and Tanganyika Two, held Italian and later German prisoners captured in North Africa. On first visiting the camps and seeing the prisoners of war, I was enormously disappointed. Having heard so much from the adults around me about how evil these people were, I had expected them to look monstrous, but I only saw people like us.

My earliest ambition was to grow up fast enough to be a tail gunner in a bomber as I wanted to be at what seemed to me the center of the action, a habit that would persist throughout my life. I was also fascinated with animals of all sorts. While living in Salisbury with my mother and sister, I kept chickens, helping the hens hatch their eggs by removing the eggs, watering the soil under the nest, then replacing the eggs so that there was enough humidity to hatch the chicks—something I learned from an elderly gardener at 14 North. I spent countless joyful hours watching those hens raise their broods.

My earliest schools in Salisbury were a disaster. I lasted one day at the convent, and was tolerated only a little longer at Blakeston, then Livingstone schools. My difficulties were, I think, due to a combination of my mother having become a very heavy drinker, which all the wartime partying and drinking did nothing to discourage, and with me becoming increasingly self-reliant and independent-minded in my childish way. At age eight I was regularly bunking school to be in the bush amongst the animals that fascinated me, when my mother, in desperation, booked me into boarding school in Bulawayo. I gave my hens to the gardener and, because I was too young to be put on the train, my mother drove the 300 miles[2] to get me there on a narrow two-strip tar road. She was an extremely proud Scot, and on that long drive she kept me entertained by reciting the poems of Robbie Burns and singing her large repertoire of Scottish songs, especially about Loch Lomond, home of the Buchanan clan since the twelfth century. When we

---

[2] English (US) measurements are used throughout this book because that is what we used in Rhodesia until 1970.

arrived at Milton Junior School[3] in Bulawayo, I refused to go in with my crying mother. We had a complete standoff, until she eventually handed me my suitcase and let me walk in on my own while she blubbed away at a distance.

I simply loved boarding school and settled into it like a duckling thrown into water. After the isolation and tension of life at home, the energy and affirmation afforded by belonging somewhere opened a whole new perspective on the world and my place in it. The headmaster, Mr. C.J. Edgecombe, was the first of several schoolmasters to whom I owe much. Over the next few years, he caned me many times but only once unjustly. On that occasion I was "beaten four" for writing on the wall by my bed in dormitory. I had not done it but knew who did. Four boys testified they had seen me doing it, however, honor being important, I would not say who had done it. When I walked out of Edgecombe's office with as much dignity as I could muster, there stood the culprit, grinning. His cheeky expression put me over the edge, and I slugged him hard in the face just as Edgecombe walked out behind me! Edgecombe swooped me back into his office and promptly gave me two more strokes: one for fighting and the other for refusing to explain why I had hit the fellow in the face. The guilty boy never did own up.

At Milton Junior I was put in the "B Stream" because of my dismal performance at the schools in Salisbury. However, once aroused by the "Milton Spirit" I rose to the top of my class and qualified for an upgrade to the "A Stream" by the end of the year. My one struggle: multiplication tables, which remain unknown territory to this day. Boys in 1B did not learn them until their second year, but those in 1A did, so I missed the lessons.

The relationships between English boys and Afrikaner[4] boys were still difficult from Boer War days, despite South Africa having come in on the English side in World War I and again in World War II. At Milton Junior, Afrikaner boys tended to get into gangs and to bully vulnerable English boys. I was one of the smallest English boys and with my notable limp, I became a natural target. The bullies grouped me with several other small boys who had obvious differences or "weaknesses." One afternoon, after a rather painful incident being beaten up by several Afrikaners, then tied up and stuffed down an antbear (aardvark) hole, I became an organizer and leader by force of circumstance. From then and into senior school, where I again found myself to be the smallest boy, I would ensure that any bully who was not dealt with by either the headmaster or prefects[5] was punished. This meant organizing the small boys into a gang; we would isolate the bully and then thrash the hell out of him with sticks. Unintentionally, I became a lifelong defender of underdogs.

To this day, the two things that most upset me are cruelty to defenseless people and to animals. Once when my pet budgie (parakeet), had a growth on its foot, I took it to a

---

[3] The combined Milton Junior and Senior school in Bulawayo, Zimbabwe, was opened on July 25, 1910, by Sir William and Lady Eveline Milton.

[4] Afrikaners are of Dutch and South African descent; languages spoken are primarily Afrikaans and English.

[5] Prefects are senior boys authorized to enforce discipline.

vet to have the growth cut off. The vet asked me to hold my budgie, which I obediently did, but when a little of its blood trickled onto my thumb, I fainted! Years later, as a game ranger and then a soldier, I ended up shedding more blood than I would have ever imagined, but wanton cruelty has never ceased to stir me profoundly.

One day in my first year at Milton Junior, while aimlessly peering out of my classroom window, I saw two gentlemen walking into the grounds. I recognized one of them as a family friend, dressed in civilian clothes. His companion, however, in khaki battledress, was a stranger. A short while later I was called out of class and surprisingly taken to the headmaster's office where he introduced me to my father to my great pleasure and excitement! My father had been sent back on a brief leave from the army in North Africa, but given the shortage of engineers in Southern Rhodesia, he didn't return. Once reunited as a family, I ceased to be a boarder and became a day boy at Milton Junior, and we all crammed into a small house about five miles out of Bulawayo.

Our house had two bedrooms. My parents had one and my sister, Ann, the other. I kept my clothes in a cupboard in Ann's room and slept on the open veranda with our dog, Mac, a mongrel born the same year as me. At night my father would wrap Mac and me in a thick camel hair blanket he had brought back from the war in the deserts of North Africa and the Middle East, where the nights often turned bitterly cold. Oh, how Mac and I loved that blanket—burrowing into it on frosty nights, watching icicles crystallize on the nearby tap. On the nights that drizzle accompanied the cold, Mac and I would scuttle into the sitting room, drape the blanket over us, and sleep together curled on the floor.

Our home was on a strip road from Bulawayo to the famous Matopos Hills[6]—hills with an immense and brooding presence that inspired one of Rudyard Kipling's famous poems and became the burial ground of choice for empire builder Cecil Rhodes, for whom the two Rhodesias were named.[7] Some days my father would drive me to and from school on the strip road. Most days, however, I rode on my bicycle. Strip roads—

---

[6] In 1840 the invading King Mzilikazi settled on the fringes of the Matopos Hills and gave them their name. Not only is the area a truly imposing region of towering granite hills, but it has numerous giant caves with walls covered in Bushmen (San) paintings thousands of years old.

[7] Cecil Rhodes (1853–1902), son of an English vicar, founded both Southern and Northern Rhodesia, which launched his dream of the Union Jack flying from the Cape to Cairo. His earliest ambition was to be a barrister or a clergyman. Ill health led him to the drier climate of South Africa where he arrived at age seventeen. After a hard beginning, he moved, penniless, to the new diamond and later gold fields, where he acquired great wealth. He later served as prime minister of South Africa's Cape Colony. Northern Rhodesia was ruled by the British government's colonial office as a British protectorate, while Rhodes' British South Africa Company financed and ran Southern Rhodesia until it became a self-governing colony in 1923. Not only was Rhodes a dreamer but he was also a man of action and courage as was demonstrated when he went unarmed and unprotected into the Matopos Hills to meet the Matabele army and its leaders to bring an end to war between them and the pioneers he was financing. Rhodes believed in the superiority of the British race, never married or lived in luxury, and died at age forty-eight. He dedicated his short life—and wealth—to expanding the empire, which he believed would be best for all people. What remained of his fortune was left to Oxford University, which established the Rhodes Scholarship, giving students from the United States, the British colonies, and Germany an opportunity to study there (Rhodes Scholarships continue to this day and have expanded to include scholars from any country). Rhodes' last words on his deathbed were apparently, "So much to do, so little done."

composed of two tracks of tar spaced the width of car wheels and often with sharp edges —could be dangerous on a bike. So, I chose to ride through the thorn bush instead, careening down a dusty ox wagon track that paralleled the strip road, startling duiker,[8] guineafowl,[9] and the occasional kudu.[10]

One afternoon, as I was bicycling home from school on the old ox wagon track, I spotted a Brownie box camera seemingly discarded under a thorn tree. I stopped and scooped it up, delighting in my "finders keepers" fortune. Back at home I was just getting familiar with my new camera when a very upset man rode up on a bicycle, demanding in Chilapalapa[11] to have his camera back, accusing me of stealing it. I of course could not deny it and hastily handed my short-lived possession over. On that day, such a deep wave of humiliation and shame washed over me that I swore to never do anything like that again.

At Milton we boys fought often. The rules of engagement were simple: If a boy challenged you to a fight, it was cowardly not to accept; whoever was challenged had the choice of wrestling or boxing; and, as soon as one boy voiced, "I give up!" the fight was over. I found myself getting into my fair share of fights, but I simply could never bring myself to give up, no matter how bloody or beaten. As such I earned a reputation for being "unbeatable." One holiday my father sent me to stay with family friends, the Napiers, on a mine near Bulawayo. Jack Napier and his wife, Ma Napier, had two sons who were in boarding school at Rhodes Estate Preparatory School (or REPS).[12] The youngest, Richard, was my age; and though the oldest, Dave, only had a couple of years on us, he was much bigger both in height and weight. No sooner did I arrive than Dave challenged me to a fight. I asked, "What are we fighting about?" because it made no sense to fight, having had no argument or difference. Dave replied that at REPS there were rumors that no boy could beat me—and he wanted to give it a go. To me this was so stupid that I floored him by simply saying, "OK, I accept your challenge, and I give up." Having just lost a fight before it ever began gave me much to reflect on; from then on, throughout my life, I only chose to "give up" if I didn't feel strongly about the issue and the importance of being right. With almost everything in life, I embraced compromise, discussion, and (reasonable) resolution.

With my father back from war, and now a "day boy" at Milton, I began to channel some of my boyhood energy into more productive endeavors. I became passionate

---

[8] A duiker is a tiny, shy, solitary antelope; the male has short, straight horns.

[9] A guineafowl is a chicken-sized, black-and-white spotted bird, endemic to Africa.

[10] A kudu is a large antelope with tawny coloring and thin, white, sparse vertical stripes; the male has large corkscrew-like horns.

[11] Chilapalapa is a language developed in the mines of South Africa that enabled the communication of mine managers and people from dozens of tribes recruited from several southern African countries. It is referred to as Fanagalo in South Africa, Chilapalapa in Rhodesia.

[12] Rhodes Estate Preparatory School, or REPS, was founded in 1932 on the model of a British public school.

about cricket. My father, who held a lifelong love of the game, coached me in every spare moment, pleased that I excelled as a wicket keeper but disappointed that he couldn't improve my batting. At age eleven, when I decided I wanted to become a poultry farmer (as I retained my love of fowls from my days back in Salisbury keeping hens), he supported me in that too. In preparation for my new "career path," I read a book on poultry husbandry, and my father encouraged me to start a small poultry business. He lent me the money to build a solid chicken house, buy feed, and select my hens and roosters. Then he showed me how to manage the bookkeeping, meticulously tracking my expenses and income (the latter primarily from selling eggs and occasionally, a chicken past laying age). This business was a valuable experience—and I was forever cured of wanting to be a poultry farmer.

My childhood ambitions were in part nipped in the bud, not intentionally, by my mother's ever-increasing consumption of alcohol, which gradually got to the point that it was almost impossible for me to avoid feeling her pain. In those days alcoholism was not viewed as a disease, but as a character flaw. My father protected my sister Ann, now a teenager, from it as much as he could, sending her to boarding school at Chaplin, a co-ed school in Gwelo (now Gweru), 150 miles away. On holidays he would whisk her away with him on his field trips. On the other hand, he didn't appear to feel the same degree of protection was needed for me, his young and independent-minded boy. Whenever I was not at school, I was inevitably at home with my mother, tiptoeing around the moments when she was completely sozzled. This created a feeling of shame that carried into my friendships; I was too embarrassed to invite friends over in case they saw her drunk as she was most afternoons. I spent my after-school hours alone, each day scouting out the bottles of brandy my mother had hidden, then pouring the smoky-smelling liquid down the sink. I struggled to understand why my mother—a loving, caring soul who had nurtured me back from the deathly grip of polio—was so weak that she couldn't quit the one habit I pleaded with her to give up.

My father eventually decided to divorce. The final straw came one morning when my mother, visibly already drunk, attempted to join my father, sister, and me having breakfast. My father was about to depart on another work trip, taking Ann as always, assuming I would remain behind to dutifully look after my mother. This time however, I could no longer bear it, and there at the table I broke down sobbing. My father, stunned, immediately assured me he would take me with him. This caused a bit of commotion in rearranging the car because there wasn't room for me having been already packed with Simon (my father's survey assistant and cook), the survey equipment, camp kit, and food. Ultimately it was decided that I could take Simon's place, agreeing to be the best survey assistant and cook I could possibly be. The three of us had a wonderful trip. Our focus was primarily helping Ted Davison, the first game warden of what is today Hwange National Park, establish dam sites for wildlife. The trip transitioned me from being my mother's guardian to my father's adventure companion, and the latter I reveled in—accompanying my father far and wide, kindling my love of the land and the wildlife that roamed it.

Two of my most vivid childhood memories traveling with my father occurred along the Zambezi River, which forms much of Southern Rhodesia's northern border. The first was one far up the river into Barotseland (Northern Rhodesia), where my father was establishing gauging stations to measure river flow. This work would later lead to building Kariba Dam, the world's first mega-dam and still its largest artificial lake by volume. Our route took us through what is today the Chobe National Park in Botswana, which was far richer in wildlife and dense with vegetation than it is now. We caught the barge at the Wenela[13] Camp on the Chobe River, which took us through the swamps to our highest point on the Zambezi in Barotseland, and on our return, we were joined by a nun from a local mission. Within a few hours of being on the barge together, I had an urgent need to pee but was certainly not going to do so in front of a nun! My father finally noticed me writhing in agony, asked what was wrong, and humored by my childhood innocence, instructed me to take care of business at the back of the barge. My face instantly reddening, I hurriedly relieved myself while the nun politely looked the other way.

The second vivid memory of "adventuring" with my father also included Oliver West, at the time the country's chief veld (range) management expert. Oliver was a close family friend, a warm genial character, and filled with good-humored pranks. He brought with him a fellow naturalist, Tom Kennon, from Matopos Research Station. The four of us traveled in three heavily laden half-ton trucks down into the middle Zambezi Valley[14] in what is the bed of the Kariba Dam today. My job was largely to keep meat stocked for our meals by shooting guineafowl and impala[15] as we traversed over more than 900 miles of primitive bush tracks. Each night by our campfire we listened to the drums of the Tonga people beating all up and down the river, either for dancing or for conveying the day's news, accompanied by roaring lions and grunting hippo. My father spun stories of his earlier survey work in the Zambezi Valley and of camp life with my mother when Ann was just a baby. In one of my favorite campfire stories he shared how families, who had only ever seen adult Whites, would come from miles up and down the river, crouched in their dugout canoes, just to get their first glimpse of a White baby. My sister Ann quickly and fondly became known as "Zambezi Ann." (When she died many

[13] Wenela is the common name for the Witwatersrand Native Labour Association (WNLA) set up by the gold miners in South Africa as a recruiting agency for migrant workers.

[14] The Zambezi Valley is named for the river that created it. It extends from the Victoria Falls gorge to where the Zambezi River enters Mozambique. The middle Zambezi Valley is now covered by Lake Kariba. The lower Zambezi Valley extends from Kariba to the Mozambique border.

[15] Impala are medium-sized antelope with slender legs, relatively long necks, and sleek, tan-colored coats. Males have slender, lyre-shaped horns.

years later, I buried her ashes on an island in Lake Kariba behind the dam near where my father's old survey camps had stood.)

When not on trips with my father, I began spending most of my school holidays with my favorite of my father's three brothers, Guy.[16] He had volunteered for the war on the same day as my father. He joined the Rhodesian gunners and pursued an active army life fighting across North Africa and up into Italy, earning a Military Cross (MC) while commanding the Rhodesian Anti-Tank Battery in the epic fight against Rommel at El Alamein. On his return he started ranching near Que Que, in the center of Southern Rhodesia, as part of an ex-serviceman's settlement scheme. He bought a ranch on terms as well as an old army truck, and shortly thereafter received his first herd of cattle on loan from the Cold Storage Commission.[17] Guy was a bachelor and gave me the utmost freedom to roam on horseback or hunt on foot over thousands of acres of bush full of sable,[18] wildebeest,[19] kudu, duiker, reedbuck,[20] leopard, and baboons. It was on Guy's ranch that I shot my first kudu on my sixteenth birthday, using his antique .303 Enfield, which had a broken extractor and required that I carry a ramrod to eject the cartridges. And it was with Guy and his neighbors that we started Rhodesia's first polocrosse club, in which we played a mounted version of lacrosse. Both children and their parents rode, while the latter consumed enormous amounts of beer.

What a character Guy was, and how I loved every moment of ranch life. I finally could invite friends over, spending hours upon hours with a group of boys who were the ranch workers' sons. Two of us "picanins," as Guy called us, became inseparable friends; Mishek and I rode and raced on donkeys bareback, trapped birds, fished, made snares out of tree bark, made fire without matches by rubbing two sticks, wrestled in the mud, and generally wore ourselves out by day's end. I eventually graduated from donkey riding to horses under Guy's tutelage, which consisted of asking me each time I fell off, "Who the hell told you to get off your horse? Get back on—now!"

Common names given by Africans[21] to their children at the time were Shilling, Sixpence, or Ticky, after the most common coins of those days. Guy's cook, Ticky, ruled the roost, telling Guy what supplies to buy in town, making the butter and biltong (cured meat), and generally fussing over us. I developed the habit of going out to hunt before dawn and getting back at eight or nine in the morning in anticipation of a delicious

---

[16] My family rarely referred to relatives by their relationship to us. Thus, Guy was always "Guy" to me, never "Uncle Guy."

[17] The Cold Storage Commission was a parastatal enterprise created in 1938 to help develop the country's beef industry.

[18] A sable is a large, brown-black antelope with both sexes having curved scimitar-like horns.

[19] Wildebeest are large, spindly-legged antelope with short, upward-curving horns, best known for their annual mass migrations in East Africa.

[20] A reedbuck is a medium-sized antelope that lives in open grasslands close to water; males have small forward-curving horns.

[21] I use the term "Africans," meaning indigenous people, although I am an African of four generations who has never seen myself as anything but African.

Ticky-conjured breakfast: mealie meal (cornmeal) porridge, eggs, and bacon. Ticky, by that time, would also have received the milk the herders had extracted from a few ranch cows and carefully separated it to get the cream for making butter. He would then leave the last of the thick cream clinging to the separator for me, which I would spoon onto my porridge with eager gratitude. Contrary to common African names, Ticky named his first two children Allan and Ann, making for an honorable connection. Little did I know at the time that once my school days were over, troubles would erupt in our country that would forever separate me from Ticky, his children, and many others.

CHAPTER 2

THE "GAME" OF LEARNING

Adolescent Lessons in Building Character and Breaking Rules

1948–1952

To hell with my ego.

Plumtree's boxing team (1948); I'm seated bottom row, first on left; my friend Bob Vaughn-Evans is seated next to me.

I turned thirteen—officially a teenager—in 1948. It was time for senior school. My father managed to get me into Plumtree, a boarding school in Southern Rhodesia on the Bechuanaland (now Botswana) border. A product of the Victorian era, Plumtree was based firmly on the English public school system where boys were expected to be hardy, self-reliant, and uncomplaining. Most of the boys came from ranches, farms, and mining towns along the rail line that extended through Bechuanaland to the Copperbelt in Northern Rhodesia. When the railway to Southern Rhodesia was built, all the sidings[1] along it from the Bechuanaland border to Bulawayo had been named after trees—Plumtree, Figtree, Gumtree, Syringa, and Marula. A school was needed somewhere on that line;

[1] Sidings are places where trains could pass each other; permanent staff were stationed at sidings to maintain sections of the line.

Plumtree was chosen because the railway ganger[2] and his wife had nine boys and girls—the most children enrolled. The school badge displayed a tree with nine plums, although by my time there were 300 boys and no longer any girls.

Academically at Plumtree I felt no passion for any subject, though I enjoyed history and loved the logic of geometry. I bounced between the top and the bottom of class, showing no constancy. One term I would be second or third and next term, second to last. Every time I sank to the bottom, my father threatened to send me to Plumtree's main rival, Milton Senior School in Bulawayo. I would vehemently promise to do better and would climb to the top again, only to once more slide to the bottom. In my weekly letters home, my father took to returning mine with all my misspelled words underlined (sometimes I'd have three different spellings of the same word in one letter!). One day, entirely by coincidence, I read an article on research into spelling. The researcher had concluded that many famous and extremely creative people were renowned for their inability to spell correctly. I eagerly referred the research to my father, who never highlighted my misspellings again.

Though Plumtree produced few outstanding scholars (certainly not me!), the teachers cultivated leadership, character, incredible pride, and teamwork. It was said that boys from Plumtree had a higher chance of selection for the Royal Military Academy Sandhurst in Britain than any other school in the realm. Many of my schoolmates did go on to Sandhurst, and more than one won the "Sword of Honor" for best cadet in his class. Looking back years later, I realized we were simply being groomed as fodder for the Empire—trained to lead, serve, and die for it—yet my experiences and friendships at Plumtree had a profound influence on me.

One such friendship, which became a lifelong one, was with Bob Vaughan-Evans. Bob and I were both assigned to live in Gaul, one of the four boarding houses at Plumtree (the boarding houses were named after Anglican bishops: Gaul, Grey, Lloyd, and Milner). I knew Bob from my younger days playing cricket against his team at REPS. Now that we were attending the same school, we became inseparable. He was exceptionally tough—built like a bull terrier—so I did my best to avoid competing against him in sports. When it came to boxing, I would spend hours breaking a sweat in the boiler room to lose weight just so I could box in a lower weight class and avoid having Bob beat the hell out of me. In rugby he was a better fly half, so I played scrumhalf or flank. In cricket, we were both on the first team but naturally played in different positions. I didn't have to worry about him in the Bisley[3] rifle shooting competition because, although a good shot, Bob didn't make the team.

My preteen reputation of not giving up in schoolyard fights stuck with me in Plumtree's sports, even when it seemed to the contrary. Shortly after I was dropped from the

[2] Railway gangers were employees stationed at regular intervals along railway lines, responsible for maintenance of the tracks and changing the points when trains had to pass one another.

[3] The original Bisley was a long-range shooting competition done on a special course that was created in Bisley, England, in Victorian times to improve the marksmanship of British soldiers. The competition was replicated in several British colonies.

school boxing team, the boy who replaced me fell ill just before an important match. I was away on a scouting camp trip at the time when out of seemingly nowhere a car arrived. Inside was Plumtree's boxing coach, who briskly swept me back to school to box. When I saw my opponent in the changing room, I nearly wet my pants. Not only did he look far stronger than me, but he also boxed under professional coaching and was the Matabeleland Champion in our weight class. In the boxing ring I resigned myself to a terrible hiding, surviving the first two rounds mainly by switching between boxing south paw and orthodox (either left or right hand extended), confusing my obviously superior opponent. At the end of the second round, I collapsed in my corner, dreading the bell for the third and final round. Amid my raspy heaves for air, I glanced down to see that the Plumtree badge on my white, sweat-soaked vest was covered in blood! The "small boy" me might have fainted, but my "never give up" spirit surged through my body. Looking up at my opponent dancing in his corner I simply thought, "No you don't you bastard, let's sort this out!" To everyone's disbelief, including mine, I won that fight. Even more amazing, aside from the team's captain, I was the only Plumtree boxer who won that night—most of my teammates took a beating against professionally coached boys.

The most exciting time of the year at Plumtree was the annual interhouse athletics competition. Taking place over a long weekend, the event lured parents in from far and wide but there was nowhere for them to stay in the village. They took over our dorms, meaning our rooms, showers, and toilets were allocated to mothers and fathers. We boys had to manage as best we could, camping and sleeping wherever we found space.

The culminating event of the competition was the interhouse tug-of-war. Each house's "tug team" consisted of seven boys. The teams were limited to a total weight of about 1,000 pounds and hence formed of boys with the best strength-to-weight ratios. Naturally, we all wanted to be selected for the tug teams! In our final year at Plumtree, both Bob and I were selected to represent Gaul's tug team. After a grueling final pull against Milner House, Gaul House won, though none of us had the strength to stand when it was over.

The incredible team spirit instilled at Plumtree carried a lesson that gave me more than one understanding of "not giving up." Because of my bout with polio, I was a slow runner. The fattest boy in school could beat me, and thus in rugby I always ended up on the second or third team. But with cricket I set my sights on being a superior wicketkeeper and playing internationally for Rhodesia and perhaps even the Springboks.[4] I studied everything I could about Sir Godfrey Evans, England's famous wicketkeeper, especially how men like him handled fast balls on the left side when standing up to the wicket. My reaction time was always very quick, and that enabled me to keep wicket for Gaul's first team.

[4] South Africa's team, the Springboks, was later renamed the Proteas.

Then a new boy arrived, Don Arnott, who was also a wicketkeeper. As I watched Don play, my heart sank. He was a natural and far better than me. My dream of playing internationally came crashing down. I was jealous, hurt, and angry. I spent one sleepless night grappling with my feelings, debating the value of an individual over the team and my place within it all. That night was a turning point in which I concluded, "To hell with my ego; my team's success was more important than me." I chose to continue playing for my house and the school's second team, while becoming increasingly proud and supportive of Don's performance as wicketkeeper. Don went on to do extremely well, playing first-class cricket for Rhodesia. In turn, I shifted my focus to other interests. In doing so, my hunting days from my time with Guy at his ranch vied for attention. But first, I would need a gun.

In Plumtree's earlier days, boys could keep guns and horses at school, but by my time both were prohibited. One day in the village's only general store I saw a Cogswell & Harrison folding .410 shotgun, priced at only five pounds because it had a broken stock. Unable to resist this deal, I decided to buy it. However, if I was going to go against the school rules, I first needed to tell my father because I would never deceive him. He was always willing to listen and gave good advice. When the headmaster, for example, informed my father that I refused to be confirmed in our Anglican school church, my father carefully listened to my reasoning (which was based on the hypocrisy I observed in devout adults). He then let the headmaster know he would not intervene; it was to be my decision. Only after I was finished speaking would my father ever share points I might not have considered, and he would always leave the final decision to me. The same happened when I wrote to him this time, telling him I wished to buy a gun and keep it at school. My father laid out the consequences if I was caught, reminding me that I would have to face them on my own. I thanked him and bought my first gun from Fifi Jones' store at the Plumtree railway station.

The proud owner of my first gun, I began almost daily to sneak off in the dark, right before each crisp African dawn, to hunt duiker and guineafowl (generally, this meant skipping our one early lesson period before breakfast, which was English, for which I have suffered since!). When successful, I took my quarry to the home of a young couple in the village. They would graciously cook it up, and I would swing by later to collect my share.

I kept my gun where least expected, under the floorboards in a prefect's study. I had only to remove or replace it by lifting the trapdoor in the floor when the occupants were asleep or away. I stowed my ammunition separately, in my own study, tucked within two hollowed-out books on my bookshelf (the books purposely selected because I warranted the reading so dull, no one would remove them). Wearing a loose jacket, my ammunition and gun fitted neatly in my clothing; once out of school grounds and into the bush, I simply strapped my gun to the frame of my bicycle.

Both my gun and ammunition were never discovered by anyone during my remaining years at Plumtree, though on one occasion, during exeat (when we could temporarily sign out from school), I came very close to being caught. On that day I was bicycling along the Bechuanaland border, intending to explore the area toward Mphoengs Border Post, which was about fifty miles away along a track through the bush. I was on a quest to find eagle nests, as eagles' eggs were especially prizeworthy in my egg collection. My gun was strapped to my bicycle in case I encountered guineafowl or duiker. Since this was my first time in the area, I almost rode right into a police outpost I didn't know existed. I cautiously skirted around the police buildings and was well past them when two police constables started blowing their whistles. I stopped and walked my bike back to them. The officer in charge was sitting in his car and told me to get in, instructing one of the constables to take my bike back to the station.

At the primitive one-room police station, the officer in charge, whose name I learned was Tucker, questioned me. He accused me of running away from school and said he would have to escort me back. I pleaded my case, "Sir, I'm simply looking for eagle nesting sites, not running away from school!" I told him I was signed out on exeat but meant to be with four other boys fishing at a dam near Plumtree. Tucker insisted on calling Plumtree's headmaster, who confirmed that I was signed out on exeat, but was enraged when he learned I was on my own so far off in the bush, demanding I return immediately. I passed up Tucker's generous offer to drive me, pointing out that I had my bike (my mind buzzing about what to do with my gun, which thankfully they hadn't yet questioned). Tucker, now more friendly knowing I wasn't lying to him, agreed to let me simply ride back to Plumtree by myself, much to my relief.

I hastily pedaled away from the police station. At the first kopje (rocky hill) I came to, I hid my gun between two rocks, covering it with leaves and wiping out my tracks. Back at school I was caned and made to sign my name on a form in the prefects' common room every hour over the next couple of weekends so they could keep an eye on me. That night, once everyone had gone to bed, I stole off and rode in the dark all the way back to the kopje that was sheltering my gun. I had both gun and ammunition safely stowed in their hiding places at school before dawn. A while later I encountered Tucker again, who had come into Plumtree's village for a fete (festival), and we had a friendly chat. I told him that he got me into trouble just as I said he would, and he apologized.

Midway through my five years at Plumtree our boarding house, Gaul, gained a new housemaster, a miserable bugger named Milward, whom we nicknamed Scab. Whereas our former housemaster had acted with dignity and honor and been especially kind to new boys, Scab spied on us, sneaking around at night and opening our study doors without knocking, hoping to catch us in devious acts. I decided to spy on Scab's spying, and after getting a sense of his "routine," I placed a large wooden trunk in a dark spot from which he often conducted his surveillance. Sure enough, Scab came sneaking along one night and unexpectedly tripped over the trunk, which sent him sprawling to the ground. Nursing both bruised shins and dignity, Scab scuttled off in the moonlight followed by our laughter. A little later I wired the outside door handle of my study. My roommate and I had ways to open and close the door carefully, but anyone grasping the handle unawares would get the shock of a 220-volt current. Sure enough, one evening we heard a tremendous shriek from Scab, who had, as usual, tried to open our door without knocking. He was furious. I apologized profusely, admitting that I had wired it in case anyone ever snooped but that "I had *never* imagined our housemaster wouldn't knock before entering." My "traps" caught Scab beautifully but did nothing to improve our relationship, as I would eventually learn—the hard way.

By tradition, new boys enrolled at Plumtree went through a two-year initiation during which each boy had to "fag" for seniors. This meant being in servitude to them, tasked with cleaning their studies, organizing their cadet kits, polishing their rugby boots, carrying their books to school, and more depending on the leniency of one's fag master. New boys were also subject to ritual harassment. When hot water ran out in the seniors' baths upstairs, which it commonly did in winter, seniors simply shouted "fag!" and the new boys had to drop everything and run outside in the freezing cold to the firewood boiler, where they would line up one by one, shivering in their shorts and stockings (no matter how cold each winter might be, and if you were a new boy, you had to wear shorts and stockings; long trousers were not allowed—even to church—until your third year). The last few boys to get in line had to hoist buckets of hot water back to the seniors' baths. Our toilets, on the other hand, were outside and comprised of partitioned box seats with open holes over buckets, which were emptied into an ox cart by its driver each morning before dawn. Even going to these outside toilets was a "character building" exercise on winter nights.

Perhaps the worst of all the ritual harassments were the dreaded marula chases. Plumtree's grounds had plenty of marula trees (*Scelerocarya birrea*) that produced an abundance of hard, green fruit (the same fruit that, when ripe, is used to make a fine cream liqueur). During marula season, each morning before breakfast the seniors would line up along the route from the boarding houses to the dining hall and call on us to run to the dining room. In turn, they would hurl marula fruits at us. As if this wasn't enough of an injustice, new boys had to collect the marula ammunition for this ritual.

In 1951 I finally was senior enough at Plumtree to be a fag master. But Scab, still Gaul's housemaster, announced that he was changing the tradition. Rather than automatically having a fag, seniors now had to be elected by the whole house to be fag masters. Scab duly assembled us all to vote, then shared that that wasn't all—there was a stipulation: boys could vote for any senior except Savory! I was automatically disqualified from the "election." In the end, Scab got back at me—and gave me my first experience with "controlled" democratic elections.

In 1952, my last year at Plumtree, Scab, to our undeniable glee, departed, replaced by a new housemaster, Cliff Coventry. At the time, Coventry, who would ultimately become a legendary Rhodesian school master, had an old T-Model Ford lorry. On long weekends when many boys went home, he would load up the remaining boys in his lorry and take us camping. On one trip I asked him if I could borrow a shotgun in the village and bring it along; he said fine. With my "borrowed" shotgun nestled at my side, I joined Coventry in the drive to our campsite: a remote dam by a granite kopje. As we were unpacking, two ducks flew up from the dam and circled above us. I quickly shot them. Coventry turned to me and with a smile remarked, "Savory, you shoot awfully well with a *borrowed* gun." From then on, I knew that Coventry knew I owned a gun; indeed, he began seeking my advice about where he might find ducks or guineafowl over weekends, and I gave him good advice. Later that year—after a hold up with a jammed .303 rifle—my final shot won the Bisley Trophy for Gaul House. Coventry was one of the first to congratulate me, giving a knowing wink and a comment about my shooting all week paying off.

In addition to hunting, while exploring the bush country surrounding Plumtree, I developed a passion for falconry. My first bird was a young female Ovambo sparrowhawk. She was very forgiving of an ignorant schoolboy who, knowing nothing about falconry, boldly robbed her from her nest to train her. Trying to learn all I could and having little access to information, I got hold of a translation of an ancient Persian treatise on falconry, the *Baz-nama-yi Nasiri*, and wrote it out entirely by hand so I had my own copy (to this day I have that handwritten version). During the day, my ever-patient sparrowhawk stayed on a perch on the lawn; at night I kept her in my washroom locker. Some fifteen years later when I visited Plumtree, I went to my old locker and found that the back of it was still white with hawk mutes (feces)!

My passion for fishing also blossomed while attending Plumtree. My most memorable catch happened when I was thirteen and while spending a school holiday on my uncle Guy's ranch. I went hunting on horseback one morning, riding along a clear stream in a vlei (meadow). Out of the corner of my eye, I caught sight of a barbel (catfish) swimming at the very bottom of the stream. Though I had no fishing gear with me, I decided I was going to catch that barbel. I tethered my horse, removed a small straight pin from my hatband and then bent the pin into the shape of a hook. Using my knife to strip the inner bark from a nearby tree, I twisted fine strands of inner bark fiber together, as Mishek and I had done in our childhood years, forming a thin fishing line about five feet in length. I managed to secure the hook to the line by wrapping it with a hair from my horse's tail. Then, after cutting a suitable stick for a fishing rod and catching a grasshopper for bait, I stalked the barbel, crawling on my hands and knees to the edge of the stream. When I was close enough, I carefully dangled the grasshopper in front of the unaware fish. Before I could blink, the barbel shot forward and gulped the grasshopper. In a matter of minutes, I had the slippery fish in my saddle bag. I rode home, triumphant.

Amid my sporting and wildlife endeavors, at age fifteen my father began pressing me for what I intended to do following school, urging me to seriously think about the future. Reality had formerly snuffed my dream of being a poultry farmer, as well as a tail gunner in a bomber, so apart from those the possibilities were endless! By this time, I was spending all the holidays I could with my uncle Guy on his ranch. I observed Guy a great deal more carefully than I had before and realized he was not much of a farmer, but he was a born leader. At farmers' meetings Guy would stay silent, long into discussions, then when he spoke everyone straightened up and listened. His words held great sway.

In a colony where the Whites were paternalistic and a few outright racist, Guy was different. During World War II, when commanding the Rhodesian gunners, Guy had befriended Neil Wilson, an English-born Rhodesian journalist, and David Stirling, founder of the SAS.[5] The two formed the Capricorn Africa Society in Southern Rhodesia in 1949. Guy was one of the first to sign on. Branches were soon opened in Kenya, Nyasaland (now Malawi), Tanganyika (now Tanzania), and England. Their Capricorn Declaration stated their beliefs: "We hold that all men, despite their varying talents, are equal in dignity before God and have a common duty towards one another . . . We hold that the differences between men, whether of creed or colour, are honourable differences."

While this credo resonated with me as a young boy, it had more to do with my admiration for my uncle than a genuine understanding of the need to state it. I was still a small fish swimming around in my fishbowl, incapable of understanding all that lay outside of it.

Listening to Guy's views on where the world was going—and reflecting deeply on my future—I concluded that a military career, though a popular choice for some of my classmates, was not for me. I felt there would be at least twenty years of peace as we had seen between the two world wars, and I did not want to spend my life as a peacetime soldier. Instead, I decided to follow my passion for wildlife and be a game ranger. I was encouraged to follow that goal by my father's close friend and my godfather, Eustace Poles.[6] Eustace had come out from England in the 1920s to join the British South Africa Police (BSAP).[7] During World War II, he fought behind Japanese lines in Burma (now Myanmar),

[5] The SAS (Special Air Service) was and is the premier special operations unit of the British Army.

[6] For a photo of my godfather and me, see the beginning of Chapter 3.

[7] The British South Africa Police (BSAP) was originally a paramilitary force formed by Cecil Rhodes' British South Africa Company, from which it took its original name, the British South Africa Company's Police. Consisting of mounted infantry, the BSAP later became the police of Rhodesia, though always known as the BSAP, never the Rhodesian Police. Officers were trained both in police duties and as soldiers (the BSAP regiment fought in both world wars) up until Rhodesia became Zimbabwe in 1980.

earning a Military Cross (MC) as second-in-command of Peacock Force, a unit in the British XIV Army headed by Lt. Colonel Edgar Peacock (an ex-Burmese forester and Rhodesian farmer).[8] After the war, Eustace joined the largest game department in Africa at the time—the Northern Rhodesian Game Department—and become one of three senior game rangers. Eustace's stories of his work as a game ranger made me eager to start.

However, there was a snag in my plan. At year's end I would be sixteen and there was an unwritten rule that no one could join an African game department under the age of twenty-five because of the responsibilities and dangers of the work. How on earth could I wait nine years? My godfather believed game departments would need professional biologists in due course and persuaded me to go to university, to not only fill the time until I was twenty-five, but to qualify to be more than a game ranger. To me, this made perfect sense.

With my mind made up on my future career, which would require a detour through university, I hastened to make sure I passed all my school subjects at Plumtree, as well as pass the Cambridge exams.[9] I did rather well in the exams due to the encouragement of my sister, Ann, who said she would buy me a watch if my scores were higher than hers (and they were). I earned a "Matric Exemption," which qualified me for university in South Africa with no further schooling necessary. (We had no Rhodesian university at the time, and a British university would have required an additional year and "A levels" to pass.) I applied to my father's alma mater, the University of Natal in South Africa, and was accepted, but at sixteen I was deemed too young and innocent for the hard-drinking postwar student life there, and hence I would have to spend another year at Plumtree. But now the question of further academic effort at Plumtree became, well, academic. With no exams I really needed to pass, and no intellectual talent or passion for anything but my future life in the bush, I had a wonderful final year at Plumtree, bunking as many classes as I could to hunt, hawk, and fish, and enjoying all the sports I loved. It was a glorious year of pure fun.

Having decided on university and a game department career, I spent the three months between leaving Plumtree and going to university with my godfather, Eustace. At his Northern Rhodesian Game Department base in Mpika, I received my first introduction to a game ranger's life. Right away I learned better ways of shooting than I had ever acquired at Plumtree, whether as part of Bisley or my own endeavors in the bush. The most critical of these skills was keeping both eyes wide open when shooting a rifle. Once I had the hang of this, Eustace gave me a 7mm Mauser rifle and a box of 1,000

---

[8] For more on Edgar Peacock, see Richard Duckett's *Jungle Warrior: Britain's Greatest SOE Commander (The Life of Lt. Col. Edgar Peacock DSO MC & Bar)* (London: Chiselbury Publishing, 2025).

[9] Cambridge exams were developed at Cambridge University and used by schools throughout the British empire to measure student ability.

rounds of ammunition to practice out on the almost-never-used airfield. In no time I was hitting a small tin can at 100 yards offhand. I could then keep hitting the can as it rolled, smashing it to bits. I also caught and properly trained my first wild brancher,[10] a sparrowhawk, followed by a goshawk; made my own falconry gear; learned what I could of gunsmithing; and more. With growing excitement, I longed for my career as a game ranger to begin.

[10] A brancher is a hawk that has just left the nest and is beginning to fly.

CHAPTER 3

THE "WORK" OF LEARNING

University Lessons in Love, Mambas, and Mud

1953–1955

Play was over, and I would have to work damned hard.

Hawking with my godfather, Eustace Poles, just prior to leaving for university (1953).

Having reached the age of seventeen, I was now considered old enough to enter the University of Natal. I duly traveled down to Pietermaritzburg, South Africa from Southern Rhodesia along with many other students on the long-haul train. Paying for university had become an issue, as my mother had thrown the family into serious debt, and my father could not afford my board and tuition. Fortunately, there was an ex-servicemen's fund to which my father successfully applied for a loan that I agreed to repay. Having enjoyed my final year at Plumtree doing little studying—and not even seeing myself as university material—this debt left me more than a bit panic stricken. I realized that play was over, and I would have to work damned hard.

Although students could apply for various grants, I didn't bother, knowing I had little hope of success and could not face the embarrassment of rejection. I was in fact raw, immature, socially awkward, shy, and naïve. I had never been in a class large enough to require tiered seating and sat far in the back at the top, feeling very out of place. The first professor to address us noted that we had arrived as useless little buggers (his words) and might leave equally useless, but while we were at university we were going to be taught how to learn! And that is indeed what I got out of this new phase of my education.

In step with becoming a game ranger, I chose to major in zoology and botany. For the first time I took my education seriously. Though I partied on weekends, memories of my alcoholic mother reminded me of what a lack of control could do to a person. I was always in control of myself, no matter my behavior, even when I might have appeared as wild and drunken as any other student.

I quickly became fond of many of our professors and lecturers. When any of us students turned in shoddy work, the professor would announce publicly that the student's work was not good enough and severely criticize it. To be clear, our professors were not *personally* criticizing us. They would emphasize that, as future scientists, we needed to expect harsh treatment and criticism of our work, but never of our motive or character. Scientific criticism was not personal criticism. Our work, not our minds, would be judged by sacred standards of academic rigor. What a refreshing idea—and yet, while this was my experience with my botany and zoology professors, I would soon discover that out in the real world, academic behavior was often the exact opposite. Anyone armed with new ideas defying deeply held beliefs could face decades of personal attacks and censure. In my own case, I would be labeled an unscrupulous heretic and banned from setting foot on university campuses in South Africa, including my alma mater, without ever hearing criticism of my work *per se*.

Academically I managed poorly in every subject but my chosen majors. I could not make head or tail of chemistry and got through with the minimum marks required by rote memorization, forgetting it all immediately after each exam. Ecology, however, struck fire. It had only recently blossomed as a field of study, stemming from the teachings of the British pioneer Charles Elton (Animal Ecology) and two Americans Frederic and Edith Clements (Plant Ecology), who in turn had stood on the shoulders of the long-forgotten polymath Alexander von Humboldt. Together they shattered the pre-Darwinian preoccupation of biologists with the classification and description of plants and animals, focusing instead on the dynamics of life altogether. New terminology flooded professional journals—words like *succession, symbiosis, co-evolution, predator-prey dynamics, climax community, food chains,* and *trophic pyramids* populated the headlines. The novelty of these concepts and the promise of making sense out of the kaleidoscope of life kept me awake, not only in class, but often half the night. I poured over journals and texts with the ardor of a romance novel addict. The material so resonated in my mind, I felt like a young Mozart, composing copious notes. I would endlessly debate my professors, who gave me high marks even when I disagreed with their teaching.

I adamantly differed with my professors in one specific area. We studied animal ecology under Professor Bush in the Zoology Department and plant ecology under Professor Beyer in the Botany Department. To study animals and plants as separate subjects was nonsensical to me; they were completely interdependent. When I questioned, "Why are we being taught animal ecology and plant ecology separately even though they did not evolve separately?" Professor Bush in zoology would sternly shut me down, stating, "Savory, if you want to be a bloody botanist go to the botany department!" Professor Beyer in botany, who used less colorful language, would then repeat the same in reverse.

Looking back, I now see that even the new language of ecology had yet to embrace concepts that were appearing in other fields: systems theory, including self-organizing systems; *cybernetics*, which deals with feedback loops; *chaos and complexity*, in which strains of order emerge from seemingly random phenomena; and what would become a lode star of my personal quest, *holism*,[1] which recognizes that complex "wholes" have characteristics neither discernible nor predictable in any of their component parts (in fact, there are no parts in nature, parts being a human construct).

I increasingly found it unreasonable to assume that humans functioned outside of the ecological principles that applied to all animals and plants, or even (later in my life) that *any* environment functioned outside these ecological principles. This revelation, however, would take time to crystalize—and I would make some costly mistakes before it did.

Another awakening in my university days came from the habit a group of us had of gathering in one of our rooms after dinner to talk about life and the meaning of it. One topic that roused me was the fact that we were at university to learn, yet society placed a higher emphasis on experience and the wisdom that comes with age. After reflecting on this for days, I exclaimed to our fledgling intellectual group, "If I have to wait until I am middle-aged or more to gain experience and wisdom, I am dooming myself to my most active years as a fool!" I boldly announced that I intended to do all I could to learn from the experience and wisdom of others *now*. In addition to everything I was reading in connection with my degree, I devoured the works of Greek and Roman philosophers and the biographies and memoirs of people regarded as high achievers. I was aided in my quest by my father's second oldest brother, my uncle Cecil (head of the accounting firm Savory & Brink in Johannesburg). Cecil, who had lost a leg in France in World War I, was a voracious reader and had a great library complete with mahogany and leather chairs. During our short university breaks I would visit Cecil and nestle into one of his overstuffed chairs, absorbing page after page of "wisdom." Ultimately, I concluded that despite the passing of thousands of years in which our material lives as humans had greatly changed, human nature and intelligence had not changed at all.

---

[1] Although J.C. Smuts published *Holism and Evolution* in 1926, its relevance was missed, and "holism" was not taught as a subject in universities.

Meanwhile, I fell in love for the first time. While standing in the queue to sign up for botany class, I noticed a stunning girl in front of me. Her dark, heavy eyebrows, luminous brown eyes, and kind face immediately had me flushed. I peered over her shoulder at her forms to get her name—Clare Tedder. From that moment I was smitten, but fortunately, for both of us, she was a good judge of character and kept me at arm's length.

Clare's father was headmaster of Kearsney College, a jacket-and-tie Methodist boys prep school in Natal. He had been friends with my aunt Nan, a teacher at a girls' school nearby in Pietermaritzburg. Aunt Nan never seemed to ruffle feathers. In my university years, whenever I visited, she quietly greeted me with a cheery disposition and a cup of tea, her grey hair neatly swept back in a tight bun. Seeing how broke I was as a university student, Nan fussed over me, even buying me a bicycle (the real intention being to encourage me to visit more often). It was only years later, from Clare, that I learned of my aunt's endeavors before my university visits, back when Gandhi was still a young lawyer in South Africa developing his philosophy of "satyagraha," or nonviolent resistance. Nan became a devotee, following Gandhi when he left for India and residing in his ashram for years. Upon her return, she promoted satyagraha to people of all races. When the South African police arrested her, she went on a hunger strike, whereupon the police broke her teeth in multiple attempts to force-feed her rather than see her die in detention. Eventually they released her, and she lived out the rest of her life as the quiet, kind aunt I loved so dearly and thought I knew. The question of why she and I never spoke of her time with Gandhi perplexes me to this day. Even in our deepest discussions about Africa, Nan never breathed a word of Gandhi. I am grateful that Clare, despite refusing my litanies of love, shared Nan's story with me.

Although both sides of my family had come from Natal, I was not really at home in South Africa and couldn't wait to leave university. I focused on learning, giving up entirely on all organized sport. To keep fit, I took to running each day. I couldn't run fast due to my polio-shortened right leg, but I could run far, which was made easy by the beautiful open country surrounding the university's men's residence (which was a dilapidated army barracks next to an unpaved airfield). My confidence in running long distances led to a memorable challenge.

A group of us students were drinking one Saturday evening in a tavern. I had just bought two Lion beers—one for myself and the other for a fellow student, Mike Creese. Mike, a bit drunk, said he did not drink Lion, only Castle beer, so I gave the beer to another student. Mike became slightly belligerent, bragging that he was a better man than I because he drank Castle. I agreed but questioned, "Everybody knows you are a better man, Mike—now how will you prove it?" Mike suggested a race. To that I responded, "I know you'll win because I'm a slow runner, but if you make the race long enough, I'll take you on." Mike suggested the fifty-six-mile course of the Comrades Marathon, famously winding through hilly country between Pietermaritzburg and Durban. Other students enthusiastically chimed in, making bets and establishing rules, starting with how we would train. It was agreed that each day for a week we would run around the block, simultaneously being fortified by the beer of our choice—Lion for me and Castle for Mike. The date was set for Friday, May 13. Our race would commence at night, following classes.

During our training week, students rallied to sign up for how many beers they would buy the winner. Clare, when I approached her to bet, said I was already too wild for beer as an incentive; she instead offered to buy the winner a box of chocolates.

Natal in May is typically cool and dry, but that season a fierce storm blew in on race night, engulfing the course with bitter cold gusts of wind and rain. Nonetheless, off we went. For a while we ran together, then after a few miles I pulled ahead, because I was freezing at Mike's pace. Friends had taken Mike's motorbike to the twenty-one-mile post and left a thermos of hot coffee for us. When I reached that mile post, cold and soaked, I anxiously grabbed the thermos, but it made a peculiar rattle. The glass lining had shattered, and the coffee was now undrinkable. I decided I had better wait for Mike, who might assume I'd broken the thermos after taking my own drink. In addition, the weather was getting all too dire for us to be running separately. When Mike caught up, he seemed to be in a bad way. I suggested looking for a farmhouse with lights on where we could get shelter, but Mike refused. Seeing no alternatives, we ran on.

After several more miles, the bitter cold seemed to be winning. Mike was moving along at an agonizing shuffle, and finally, in desperation to stay warm, I told him that I had to speed up. I promised Mike I would wait for him fifty yards from the agreed finish line; we would cross it together. From there on I managed to ward off the cold by running as fast as I could. When I reached our waiting point, I continued running in circles to stay warm. It seemed like hours (and possibly was!) before Mike appeared, just at the crack of dawn. Together we walked over the finish line, making the race a draw.

Though my long runs in the countryside helped, I deeply missed being in the bush with all its varied smells and sounds. As a result, I tried to fill every long university break (a month in July and three months over Christmas) with an adventure in the bush. In my first year, the Southern Rhodesia Forestry Commission offered me a temporary job assisting their forester, Blake Goldsmith, on the Gwampa Forest Reserve north of Bulawayo. The pay was thirty pounds a month—a healthy wage for an impoverished student. I signed on.

It had been an exceptionally rainy season when I arrived at Blake and his wife Gladys' home in the Gwampa Forest. After issuing me a blanket, tarpaulin, camp stretcher, paraffin (kerosene) lamp, and a couple of pots, Blake drove me into the forest to begin surveying timber. To survey timber, a group of men had to walk along previously pegged lines, typically ten miles or more a day, dragging a length of chain to accurately measure the distance, and gathering data on every tree: its species, breast-height diameter, and general health. This had to be done for every tree within eleven yards of both sides of the line. Men with calipers measured each tree's diameter, followed by two men dragging the chain. My role was to record the species of trees and their breast-height diameters using a pencil and paper attached to a clipboard. Every day we trudged along the designated sampling area, steadily completing our tasks. When the rain would come pelting down, we all crouched low to avoid being struck by lightning. I would cover my ever-important clipboard with a scrap of canvas. At the end of our day's pegged line, usually reached by late afternoon, we then had to walk back through the Kalahari sand forest, sheets of rain drenching us.

It wasn't long before I encountered my first real danger in surveying timber. The man in front of me unknowingly stepped over a large black mamba, the most poisonous and fastest striking of all our snakes in southern Africa. The snake reared in fright. Everyone froze. Being next in line, I clutched my clipboard below the mamba's unwavering head, feeling its beady eyes burning into me. More than six feet of its body was raised off the ground and ready to strike. Given how speedily mambas strike, I knew that if I moved, I had no hope. The snake and I stared each other down, still as stone, for what felt like hours. Finally, the mamba flipped over backward and slithered off, to my great relief. I quickly learned that the Gwampa Forest was full of mambas, and just days after my first encounter I accidentally drove over one's tail. The snake struck instantly, ricocheting off my vehicle's window frame, just centimeters from my head.

Due to flooded rivers, we could not always get supplies of food for all the men and their families, so I got the added job (to my delight) of stocking our food supply by hunting the plentiful kudu antelope. Blake loaned me his ex-army .303 Enfield, and I finessed my tracking and shooting skills to ensure enough meat for all the staff.

Midway through my stint, Blake suggested I take a couple of days off—perhaps making a visit to Bulawayo for a much-deserved break. Using a crank phone that only operated on a party line, I called some of my friends and arranged a good time in town. Blake loaned me his Land Rover and I headed into Bulawayo, donning my one good white shirt and longs (trousers). En route, along an extremely wet and muddied stretch of road, I came upon a Ford truck, its wheels sunk in the mud, stuck. A man was bent over the tracks, shoving leaves under the back wheels while an elderly couple huddled in the truck's cab, periodically revving the engine while the man pushed. Each time they hit the accelerator, the wheels spun and splattered the man with mud. Yet the truck barely budged. Seeing how desperate the situation was, I offered to help. The couple was indeed grateful, saying it took two hours to move the truck 100 yards! I maneuvered Blake's Land Rover around the Ford, backed up, and pulled it out. From there on, the road only seemed to get worse, so I stayed close, pulling their truck out repeatedly. By the time we got to the (much improved) road at Turk Mine, about halfway to Bulawayo, it was nearly dark. My day was gone, and my white shirt was now caked in mud. Resigning myself to no break after all, I turned around and drove back to camp. I forgot the incident, but more than twenty years later my father shared a letter with me that he had kept. The letter was from the couple; they had somehow learned my name and tracked down my father to express their gratitude to him.

I was sorry when my three months' time with Blake was over. In my time with him I not only learned a great deal about surveying but also about field botany (Blake was recognized as Rhodesia's foremost field botanist, despite his lack of a degree. During the war he had served under my uncle Guy and at war's end could not afford university). In addition to surveying, I had started collecting plants under his tutelage, adding them to the plant-drying presses littering Blake and Gladys' kitchen. Blake, a self-taught scientist, was meticulous in identifying the plants, researching their origins, and noting his observations. From him I learned more about the discipline and ethics of science than I ever did at university.

I spent my remaining extended university breaks, amounting to four months each year, back in Northern Rhodesia with my godfather who continued to inspire me to be a game ranger. Through Eustace I got to know another senior game ranger, Len Vaughan, and the director, T.G.C. Vaughan-Jones, or VJ. My months living and working with this team were joyous. I was eager to learn, to get a taste of the life I hoped to live as a game ranger myself, and to do any job they asked me to do. At one point, VJ unofficially issued me a new .375 Holland & Holland rifle which I would return to the armory each time I went back to university. To help me gain experience, Len let me loose near Namwala, where I shot my first lion and buffalo. The following year, while with Eustace, we crept up on a herd of elephants. Eustace intended to shoot one bull, but instead nearly bumped into a bull elephant hidden entirely in the dense bush. I was right behind Eustace and quickly grasped his shoulder, jerking him back, his body nearly brushing the bull's trunk. The bull startled and stampeded off. Eustace, without giving me time for my nerves to get the best of me, shouted, "Go for it!" I ran after the herd, managing to shoot my first bull elephant. To commemorate the event, Eustace blooded me on the forehead with its tail and later presented me with an engraved hunting horn.

Having now shot my first lion, buffalo, and elephant, I spent hours in the game department's armory inspecting every rifle, especially those that had been returned broken, jammed, or otherwise malfunctional. I wanted my own rifle, but it couldn't just be any rifle—I planned to custom-order it from Holland & Holland. A student designing and ordering a rifle from such a famous gunmaker was clearly unusual. This was reinforced some thirty years later when I was admiring a display of Holland & Holland rifles in their shop in London. A gentleman working there, dressed in a three-piece suit, struck up a conversation. When he asked my name, to my surprise, he responded, "Ah, yes, we made a rifle for you some time ago, let me go look it up!" A few minutes later he appeared with a ledger from 1955, flipping to the entry: "7mm Mauser action, hi-chro steel barrel, No. 2555, sent to C.R. Savory at Chilanga, Northern Rhodesia via Hunting Clan Transport."

With a little game ranger experience now under my belt, I traveled to Pretoria, South Africa, for an interview at the British Embassy (the game department, being in Northern Rhodesia, was under colonial rule). To my delight, I was offered the job of game ranger, provided I earned my university degree. The fact that I didn't yet meet the minimum age of twenty-five was waived, thanks to support from those in the department who knew me and treated me like a son. In addition, VJ, the director, had recently voiced the need for African game departments to employ trained biologists.

In 1956, during my final months at university, I fell in love again, this time with Shirley Garbutt, one of Clare's best friends. Shirley was full of fun, attractive, and skillfully practical, having grown up on a farm. Most importantly, she loved spending time in the bush. We soon started dreaming of a future together. To make this happen we would first have to spend time apart; I would start my job as a game ranger while Shirley remained at university to get her master's degree. Since Shirley was the daughter of farmers who could not afford further education, I agreed to pay for Shirley's continued studies while also repaying the loan from my father. I had always had a deep but unspoken desire to marry early and have a son to share my life, hence we planned to marry in a year, making a great team devoted to saving Africa's wildlife.

Final university exams came and went. I was confident that I had passed and quickly said my goodbyes, assuring Shirley that our lives together were just beginning. I got on the first train to Salisbury and never did go back to the University of Natal, even for graduation. My professors wrote to my father urging him to persuade me to return, apparently considering me a student with high academic potential.

PART II NORTHERN RHODESIAN GAME DEPARTMENT

CHAPTER 4

TO BE A DEVOTEE OR A GUNGA DIN

My Early Disillusionments

1956

They were all in shock and even those uninjured seemed paralyzed where they landed, incapable of action—something I had not seen before but would again.

Elephant control guard Sondashi Kazau and I with the Ndola "cattle killers."

Since the days of the earliest British adventurers in Africa, the bush (or the *veld*, as we called it) and its magnificent animals had both beckoned and challenged young men. Rangers anchored their identity in the teeming savannas of Africa like the Swiss clung to the Alps and the Tartars claimed blood kinship with the steppe.[1] I felt privileged to be selected to be a game ranger, and I was proud to be one of the men devoted to saving it all!

Excitement coursed through me as I jumped on the train from university to Salisbury, where I briefly stopped to spend Christmas and New Year with my father. He had

[1] Steppe refers to the dry, grassy plains of Eurasia.

moved into a senior position in the Department of Irrigation[2] and, by then, he had also remarried, having reconnected with an old friend, Thelma Marcus. Thelma was a chartered accountant (whose skills would later prove beneficial in getting my father out of debt). After a pleasant visit, I packed my binoculars and what few clothes I had in my suitcase. I was off to the Northern Rhodesian Game Department to start my new job—Allan Savory, game ranger!

Northern Rhodesia, the largest country in the newly formed Federation,[3] had become a British Protectorate eleven years before my birth and was at the time divided by the Colonial Office in London into six large administrative provinces: Northern, North-Western, Western, Southern, Central, and Eastern. The game department's headquarters were in the Central Province, which was the department responsible for game management as well as fisheries and tsetse fly control over the entire country, including game reserves and future national parks.

Getting to the headquarters in Chilanga, a small settlement a few miles outside Lusaka, entailed hitchhiking 300 miles. My father, Thelma, and my sister Ann were on their way to see friends near Banket—about forty miles north of Salisbury—and dropped me off there on the two-strip road to Northern Rhodesia. My father loaned me ten pounds, which I promised to repay with my first month's salary, and I started walking. In those days it was rather easy to hitchhike, and after catching three uneventful lifts I arrived in Chilanga at 4:30 that afternoon.

The headquarters of the Northern Rhodesian Game Department consisted of an office block (a row of rooms opening onto a wide veranda), a few houses, and a game department garden that helped feed the families. Having already visited for three months while on a previous Christmas break, I was familiar with the place and, when finding no one at the office, walked to the home of Major Ian Grimwood, who I had come to know in previous years while visiting my godfather. Major Grimwood, along with his wife, Peggy, put me up for the night. Over dinner I learned that circumstances had changed. The director, my mentor VJ, had left the department. A man named Fip Parnell was taking over as acting director with Major Grimwood serving under him. I was to stay in the former director VJ's now-vacant house until other accommodations could be found for me, and I was to report to Major Gerry Taylor, Provincial Game Officer in charge of Central Province.

VJ was one of the main voices advocating for me to start at age twenty versus the standard ranger age of twenty-five. The original plan had been for me to spend my first

---

[2] Although it was called the Department of Irrigation, it was in fact responsible for all water development and building of dams in the country.

[3] The Central African Federation, also known as the Federation of the Rhodesias and Nyasaland, was commonly referred to as the Federation. It consisted of Southern Rhodesia (renamed Rhodesia in 1964, then Zimbabwe in 1980), Northern Rhodesia (renamed Zambia in 1964), and Nyasaland (renamed Malawi in 1964). It was formed in 1953 and dissolved on December 31, 1963.

six months learning the ropes, serving under three senior rangers, two of them ones I'd already worked with—Len Vaughn and my godfather, Eustace—and the third a man by the name of Norman Carr. In addition to gaining game ranger experience under these men, VJ believed I could also begin using my university learning as an ecologist. Now, I realized that the implications and possible troubles of working under different leadership might indeed change that trajectory, but I wouldn't let that get me down. I was elated; I was finally joining Africa's largest (and best organized) game department.

To get a sense of where this placed me within the organization: the Northern Rhodesian Game Department was dominated by ex-army personnel. Overseeing the department was the director and assistant director, both at the headquarters in Chilanga. Also at headquarters was the chief entomologist in charge of tsetse fly research and control and the chief fishery officer, as well as an armorer, several clerks, and various administrative personnel.

Each of the six provinces had its own branch of the game department; each branch was led by a provincial game officer (PGO) who reported to the department's director. Each PGO oversaw all the staff and activities within their province. Each PGO had a secretary, clerks, and driver, as well as game scouts, game guards, and elephant control guards. Some also had fishery and tsetse officers reporting to them. In each province there were several game rangers stationed throughout, each having their own clerks, a driver, game scouts, elephant control guards, and game guards. Three of the provinces, which contained major game reserves, also had senior game rangers (in the Northern Province one of these senior rangers was my godfather, Eustace). All game rangers, fishery officers, and tsetse officers, as well as the secretaries of PGOs, were White. Most were also English with a few "colonial" game rangers, including me. On the other hand, most staff members were Black, with generally only clerks and game scouts having at least a school leaver (high school) education. Elephant control guards were generally older, more experienced men of great character, capable of handling heavy rifles. Game guards were generally younger and capable of handling single-barreled shotguns.

January 3, 1956, my first full day in game ranger paradise, was spent filling in endless forms, my hands covered in carbon from leaning on the many carbon copies. I was then issued a rickety three-quarter-ton Chevrolet truck, Hounsfield camp bed, canvas tarpaulin, Bialaddin paraffin lamp, and portable canvas bath and basin.[4] Of most importance, I finally signed for the Holland & Holland .375 magnum rifle I had been "unofficially" issued as a visiting student (having shot my first lion, buffalo, and elephant with it during university breaks).

The responsibilities of a game ranger were simple: deal with individual problem animals—lion, leopard, hippo, bushpig, buffalo, rhino, elephant, baboons—generally

---

[4] A portable canvas bath and basin had a lightweight folding metal and wood frame that, when upright, held a waterproof canvas wash basin and when erected horizontally held a waterproof canvas bath about six inches deep.

by shooting them; catch poachers, which involved tracking them down or ambushing them; and oversee the burning of grass and bush in the interest of re-greening the land and providing fresh feed for the game.

After my first eight days of largely mundane tasks that made me restless, a report came in of a leopard killing calves on a nearby farm. Maurice Curley, the more experienced game ranger of the two of us, was away, so Major Taylor (my newly appointed boss who I hadn't yet spent much time with) had to deal with it. He asked me if I would accompany him; I was pleased to have the opportunity to get to know Major Taylor and immediately said yes. We piled into his Land Rover and headed to the farm, arriving late in the afternoon. Major Taylor explained the plan: We would tie a calf to a tree across the ditch from us, sit in our vehicle until the leopard snatched the bawling calf, then switch on the lights, open our doors, and shoot it! I quietly thought this was going to have to be a very stupid leopard if we were to succeed.

Dark descended quickly, accompanied by rain. Minutes, then hours went by. Major Taylor frequently swigged whisky from a flask. The calf stopped bawling, settling in under the tree to sleep, followed shortly by Major Taylor, snoring. Periodically a flash of lightning illuminated the clearing across the ditch in front of us, assuring me that the calf was there under the tree, still safe. Finally, Major Taylor woke up and I said that I didn't think the leopard would show at this late hour. He mumbled agreement, started the Land Rover, and began to back it out of position. I asked what he intended to do with the calf. He said we could not drive through the ditch now, which was full of wet mud, so we would leave it. I agreed that driving through the ditch wasn't possible, suggesting instead that we just lead the calf back to its mother. Things got a bit awkward then as it became clear that the major was afraid to venture out to the calf in the dark. Exasperated, I opened my door, stepped out, climbed through the ditch, hauled the calf back to its mother in a nearby kraal (enclosure), and returned to the passenger seat, unharmed. Not a word was spoken on the drive back nor when we parted.

From that point, things went downhill between my superior and me, perhaps as much my fault as his. Twenty-three days passed in which I sat in the office, doing mundane tasks and essentially becoming Major Taylor's "Gunga Din" (Rudyard Kipling's caricature of a regimental water boy), taking trips to Lusaka to fetch supplies from government stores. When Major Taylor finally gave me a solo assignment in the bush, it ended up being lackluster—I was sent to the junction of the Kafue and Zambezi rivers to deal with a problem lion turned man-eater. It turned out the lion was old and almost toothless, and the local men had killed it with their spears before I even arrived. So, instead I spent a few hours fishing for tigerfish in the Zambezi.

The other game ranger under Major Taylor, Maurice Curley, took pity on me and persuaded Major Taylor to let me deal with problem bushpigs on commercial farms south of Lusaka near Monze. It was on this assignment that I met Game Ranger Frank Ansell. Frank, like me, was my godfather Eustace's protégé. An Englishman, Frank had fought alongside Eustace in the Peacock Force of the British XIV Army. Frank was so passionate about studying mammals that he was said to have written and smuggled out a paper on the status of the Burmese rhino while fighting behind the Japanese lines. At war's end, he followed my godfather to the game department, where he was now also a collector of mammals for the British Natural History Museum.

Frank had also been a fox hunter in England, and he ran a first-class pack of hounds trained to track down problem bushpig and leopard. Frank gave me his second team—a pack of mongrels and misfits that he had "rescued" from various places. Few animals are as cunning or ferocious as an African bushpig. On my worst day, six of the dogs Frank lent me were badly mauled, whereas I only killed two bushpigs. Not a good ratio. Frank, on the other hand, was far more successful with his well-bred professional team.

My next assignment, also thanks to Curley, was a foot tour, or ulendo. I was to report on poaching, the presence of tsetse fly,[5] and game populations in the hills west of the lower Luangwa River down to the Zambezi. It was the heart of the rainy season and thus the worst time of year for the task, but I was just glad to be out and about. Curley gave a couple of game guards and me a ride to the road to the Eastern Province. With a little effort, we recruited carriers[6] from nearby to help with our belongings and set off.

The rain was relentless and by the fourth day in the bush, spirits were low. Pindigani, our local guide, seemed to be confused and, after I had words with him, admitted he was lost. We had neither map nor compass and never saw the sun in the almost constant rain, so I followed my instinct on what direction was south and headed us there. That decision immersed us in frightfully hilly country. The carriers, who had lived in the area all their lives, said I was going the wrong way; however, because they couldn't seem to agree on an alternate route, I kept going in the direction I thought right. I shot a warthog, and we did our best to smoke and dry the meat, our other option being sopping sour mielie meal. We crossed with difficulty what could only have been the Rufunsa River, entering dense bush with many signs of buffalo and elephant. On our sixth day we climbed a high hill and caught a glimpse of the Zambezi River, the shining stretch of water renewing our sodden spirits. Eight days later, we arrived at Feira Post at the river junction, bringing our fourteen-day foot tour to a much-needed end.

---

[5] The tsetse fly carries parasites of the genus *trypanosoma,* which infect domestic animals, especially dogs, horses, and cattle, usually killing them. Rarely, the tsetse fly also carries the parasite that causes human sleeping sickness, which is commonly fatal. I share more about this fly and the government's efforts to eradicate it in Chapter 11.

[6] For centuries in Africa, much work was done on foot and that entailed men walking with loads of food, tents, or whatever was needed. These men were called carriers. A carrier commonly carried a forty-pound load to which he tied his own bedding and so forth.

A month later, my godfather wrote to me, cautioning me that rumor had it I was gaining a reputation for "bumptiousness" in the department. This was hurtful in the extreme. I had all my life (and still do) battled with shyness, and I was by nature a quiet person, but bumptious I was not. Thankfully, Frank must have worked some magic as, shortly thereafter at the end of a training course, he pulled me aside to share that I was being transferred to the North-Western Province to work under him from Ndola (where Frank had become Acting PGO). I couldn't pack my belongings fast enough! A couple of days later I drove a five-ton lorry up to Ndola where I welcomed my new living quarters: rondavels (round, mud huts) that Senior Ranger Norman Carr had built on a small plot of land he owned in a miombo[7] forest ten miles from town.

Up to that point, I had been given mundane, almost pointless tasks under Major Taylor, but now, Frank kept me busy with serious assignments. He assigned me to various game-surveying foot tours while simultaneously collecting mammals of every species I could find for the British Natural History Museum. Mammal collecting involved trapping small mammals, shooting larger ones, taking measurements, skinning each, and preserving them.

One of the most memorable assignments was placing permanent stone beacons to demarcate the northern boundary of the Lunga Game Reserve near Solwezi. While there, I was also to conduct a foot tour collecting flying squirrels and specimens of both the smallest and largest duiker species—the blue duiker and the yellow-backed duiker, respectively. Frank provided me with a team, including an exceptional game guard named William who was Frank's main trapper and skinner and who freely shared his immense amount of bush knowledge.

I duly put stone beacons where required on the boundary of Lunga Game Reserve and camped nearby. William's superb skills made it easy to collect flying squirrels, klipspringer,[8] and various small mammals by trapping them. Sighting the blue and yellow-backed duiker, on the other hand, wasn't as easy. We found countless tracks of both species of duiker, and I even became quite skilled at trying to lure them in by calling them as William had previously taught me to do—place two fingers of your right hand up your nostrils and the palm of the same hand under your chin, then wobble your hand while bleating nasally and varying the tone for whatever animal is being called—duiker, civet,[9] leopard, kudu, bushbuck,[10] and so on.

---

[7] Miombo woodlands is a vast ecoregion in Central Africa dominated by Brachystegia tree species such as *Brachystegia Spiciformis* (msasa).

[8] The klipspringer is a small, greyish-tan antelope with specialized hooves and excellent vision for navigating steep cliffs (its name in Afrikaans means "rock jumper"). Males (and some females) have small, spiky horns.

[9] The African civet is a medium-sized, catlike omnivore known for its distinctive black-and-white markings resembling a raccoon mask. It also has a short mane along its back.

[10] A bushbuck is a shy, elusive, medium-sized antelope with a distinctive brown coat of white stripes and spots. Males have short, straight horns and can be aggressive if cornered or wounded.

We bleated for what seemed like hours, but it wasn't until days later that we successfully called up two blue duiker and an impressive yellow-backed duiker. I shot them using my silenced .22 rifle. The horns of the yellow-backed duiker turned out to be exceptionally large; Frank had the duiker registered in Rowland Ward's *Records of Big Game*[11] under my name, C.R. Savory, where it remains to this day.

While on this assignment, we learned that a nearby village had reported elephants raiding their croplands. Without hesitation I assured the villagers I would handle it. In those days, it was customary for game rangers to take out a hunting license every year for up to four elephant bulls, the idea being that we were controlling elephant populations anyway and by getting a license we could get on with our work and not take leave to go hunting privately. At year's end, each ranger then typically kept their four largest pairs of tusks. I accordingly had taken out an elephant license hoping that if I shot one on control it would help repay my debt to my father and cover my girlfriend Shirley's tuition.

Within hours the elephant control guard, Sondashi Kazau, and I had tracked the elephants from the croplands down through mavunda woodlands.[12] It was almost impossible to see any tusks, the trees and brush being so dense, but thanks to a steady wind I got about three paces off on the side of one bull, undetected. I shot him as cleanly as I could, acquiring my first set of tusks, which I later sold, as we did in those days, by auction in London.

On our way back from this trip, after dropping William off at his village for a brief break, we stopped in Solwezi to spend the night in the government rest camp. Sondashi was siphoning petrol from a drum into the fuel tank on our lorry when suddenly, to my horror, the lorry lit up in flames. Sondashi immediately started to throw supplies off the truck to keep them from burning. Amidst the hail of supplies, a bundle of blankets landed at my feet. Tearing off the straps, I quickly threw the blankets over me and, now a human fire extinguisher of sorts, smothered the flames at the source—the overflowing tank. Others rushed to the rescue, throwing sand on the remaining hot spots. The tank did not explode, as it was full and overflowing. Why the drum above did not explode remains a mystery.[13] Apparently, one of the camp staff had walked by with a lit paraffin lamp that ignited the fumes. Though my blanket strategy had been effective in putting out the fire and saving our lorry, I did not come away unscathed. My eyebrows, beard, and much of the hair on my head and arms was singed if not entirely gone, and the burns on my face were so painful I didn't sleep at all that night.

---

[11] *Rowland Ward's Records of Big Game* was established in 1870 to record the measurements of game animals taken in fair-chase hunting and to share them as a matter of interest to international sportsmen.

[12] Mavunda woodlands are dense evergreen forests dominated by the tree genus *Cryptosepalum*, known locally as "mavunda."

[13] Petrol will burn, but if there is no petrol vapor (because in this case the tank was full and overflowing as it heated up) it will not explode. The drum above should have had vapor near the top and exploded, but thankfully it did not.

At 4 a.m. I woke everyone, we loaded up our lorry (which had miraculously survived incineration) and continued onward. I drove, with Sondashi in the passenger seat and the rest of the crew piled into the back. We bumped along quietly, contemplating how silent the bush was during the day compared to the long nights. During the nights, especially when moonlit, the bush reverberated with the beating of village drums, the roaring of lions, snorting of hippo, and trumpeting of elephants, all of which, to this day, move me more than the finest orchestra ever could.

I was a bit out of my senses from the previous day's fire and the painfully sleepless night so what happened next, I can't explain. My mind must have drifted, and my foot did the same on the accelerator. I had the lorry going way too fast coming round a bend and it skidded sideways on the loose gravel, hit the bank, then rolled over three or four times. The men on the back were all thrown off in the first roll but Sondashi and I, trapped in the cab, were tumbled around like clothes in a washing machine. When the bouncing and crunching stopped, I was wedged on the floorboards of the passenger side where Sondashi's feet had been, and Sondashi was gone. As I tried to calm my breathing and think, I heard Sondashi wailing for help from somewhere under the vehicle. I pulled myself out of a broken window to find Sondashi pinned under the truck. I could not get him out and had no idea of the extent of his injuries. He kept crying "Bwana, Bwana, Bwana" meaning boss, told me he was dying, and asked me to take care of his wife. Just then a group of women came to my assistance; they were from a nearby village, had heard the crash, and hurried over. Together we pried Sondashi free. His back, legs, and arms were in pain, but we could not find anything broken. (I later learned I had broken a few ribs, but at the time the rush of adrenaline seemed to mask—or perhaps lessen—my pain.) We gathered up the rest of the crew who were strewn along the road, all with minor injuries but for one who was badly scalped. They were all in shock and even those uninjured seemed paralyzed where they landed, incapable of action—something I had not seen before but would again.

We settled on the roadside in a circle. I made a fire to warm the men and in part to give myself time to think. Our "miracle" lorry, having survived flames, was now beyond recoverable—a crumpled mess of metal before us. Within minutes a van came along, and we piled those who were severely injured inside, the driver assuring me he would go straight to the hospital in Chingola. Various other cars drove by, and when the next one stopped, the driver, after confirming the rest of us were okay, seemed most concerned with the lorry. He asked if I had the logbook, which we managed to dig out of the glove box. After closely inspecting the damage, he introduced himself. Coincidentally, he was Senior Inspector of the Public Works Department, the department responsible for servicing all government vehicles. He asked what had happened, and I told him I was driving too fast, came around the corner, and the lorry skidded and rolled. I prepared for the worst. Surprisingly, the inspector simply thanked me, said he would be submitting an official report, and declared that I would not be held responsible. Apparently, although the lorry had been recently serviced, the U-bolts and shackle hanger bolts of both front springs were very loose, the bolt holes worn into an oblong shape. The accident, according to the inspector, was inevitable. Yet I knew damn well that I had been driving too fast and was thankful I had not killed anyone.

Those of us remaining eventually got a lift and took with us the most coveted of our gear: boxes of mammal skins, crates of skulls, and of course, my tusks. Back in Solwezi I sent a telegram to Frank, apprising him of the accident. When I arrived in Ndola after hitching a lift, Frank and his wife Mary invited me into their home for dinner. As we ate, I anxiously waited for Frank to say something, expecting a dressing down. Finally, I asked if he had received my telegram. "No," he responded. "What was it about?" I then confessed to driving too fast and writing off the truck, but thankfully not killing anyone. "Oh, don't worry about that," Frank laughed. "I've written off more than one!" That was the end of it, and despite Frank's terrible but seemingly excusable driving record, I made an oath to myself that this was an end to my reckless driving.

Just as I had learned so much from Blake Goldsmith in the Gwampa Forest, in my time under Frank I learned more about mammalogy and the discipline of science than I ever learned at university. In these self-taught men I found an amazing determination, discipline, and ethical scientific behavior. Yet I was becoming disillusioned by the Northern Rhodesian Game Department, primarily the higher-ups. I had arrived in Chilanga feeling excited and honored to be joining a team of men who, in my mind, were living my dream of saving wildlife and serving a cause. However, much of what I was observing, beginning with Major Taylor's indolence, had me feeling quite differently.

In one such instance, our Acting Director Parnell circulated a report exemplifying what he expected from us. He presented it as a model of excellence, saying it was submitted by a game ranger who went to great lengths describing a foot tour of the area around Lunga Game Reserve. The document, Parnell emphasized, was an example of how we should all be working and reporting.

When I made my own visit to Lunga, I walked many of the areas, visited the same villages, and saw much of the game that the report described. I also encountered the place where the author of the report had first camped. I found no sign, however, of camping on the other sites described within the report. Upon asking the local villagers, they appeared puzzled. The villagers confirmed instead that our "exceptional" game ranger had parked his vehicle at the first place, set up a large tent, and spent days on end drinking. It seemed rum, whisky, and a dash of writing talent were really what was behind the report. I'm sure it did nothing for my "bumptiousness" reputation when I submitted *my* report, revealing that Parnell's "exemplar" document was phony.

No sooner was I back from our eventful trip when Frank sent Sondashi and me off to a nearby farm where lions had killed some cattle. This time, in addition to my usual supplies, I packed a "gift" I had just received via mail from my godfather Eustace. It was his favorite rifle—a double-barreled Hollis, Bentley & Playfair .470. He had given it to me for my twenty-first birthday, which wasn't until September, out of concern that I was hunting problem elephants, often in dense bush, with only my .375 magnum rifle. I knew the .470 well, having cleaned it many times, and I wept to think that I would ever deserve such a gift (to this day, I still refer to the rifle as "The Baby," Eustace's nickname for the beloved gun). The rifle rode with me up front in Frank's Land Rover, and we arrived at sundown to find the remaining cattle bunched in a small brushwood kraal (enclosure).

Upon approaching the kraal on foot, I startled a lion that rushed off into the evening shadows. Neither Sondashi nor I had yet recovered from our lorry accident and didn't care to follow that lion in the fading light, so we decided to stay and try at dawn. Sondashi slept in the Land Rover as he was in the worst condition, and I slept wrapped in a blanket on the ground, against the kraal's brush fence, just in case the lions returned. I slept like a log, waking at dawn to find a half-eaten cow about ten paces from me! The lions had returned, crept into the kraal, killed another cow, jumped over the brush fence with it, and feasted on it—all while I blissfully snoozed. Realizing we were not in a condition to track as usual, Sondashi and I decided to use poison, which to me was both unsporting and cowardly, but I didn't see another option. We rolled the dead cow over, placing arsenic powder into cuts I made into the meat, being careful not to leave any trace on the hide. As an afterthought, I also placed some of the arsenic into cuts on the cow's nearby fetus (unborn calf), which the lions had torn from her and abandoned. We then spent the day fending the vultures off our "bait."

That night the lions returned; we let them go about their feeding. At dawn we found they had rolled the cow over and eaten the other side—leaving the poisoned side untouched! However, the fetus was gone. Having no option now, we took up the tracks, ill and in pain as we were, followed closely by a group of villagers with spears and dogs. Soon we came upon a dog that was shivering and arching in pain; it died in front of us, and we concluded it had licked some of a poisoned lion's vomit. We urged the villagers to take all their dogs away. Shortly after, we came upon a lion lying down, almost dead. With one shot I put him out of his misery. Commotion erupted amongst the villagers, who descended upon the lion with spears, seeking to pull out his mane and claws. I had to intervene, first unloading the barrels of my .470 to avoid an accidental discharge while restoring order.

When calm had returned to the group of villagers, we left the dead lion and one man behind to guard it. Sondashi and I continued tracking the second one, which led us into a dense reed bed. We circled around the whole reed bed, finding no exit tracks. Then suddenly we heard moans from deep inside the reeds. We quickly took up the track where the lion had entered, the pushed-over reeds making it easy to follow. I took the lead, primarily because I could not have anyone tracking in front of me if I had to shoot in such dense vegetation. I instructed Sondashi, who followed with a 12-bore shotgun, to remain well back. Eventually I made out part of the lion's hindlegs and quietly raised my rifle, trying to assess where his chest might be. Before I could pull the trigger, however, the lion charged. I quickly fired, but to my horror heard only a hollow click. I had forgotten to reload my rifle after dealing with the excited villagers! Thankfully, Sondashi had disobeyed me and was only about three paces behind me. He fired his 12-bore shotgun past my left side, the large buckshot pellets knocking the lion slightly off balance. That gave me just enough time to chamber one round and fire. What a close call! And what a lesson. Never again would I go into action without checking, checking, and rechecking my gun.

In July, Shirley came up to visit from Natal, where she was still at university. We had long discussions about our future. I shared my disillusionment in working for the Northern Rhodesian Game Department. The team had good rangers and people whom I deeply respected, but they were a minority. The two men now directing the department regarded their time there as just an exotic career without any long-term commitment to the country, people, or wildlife. I decided I wanted to transfer to the newly forming Southern Rhodesian Game Department, thinking (in my naïveté) that we Southern Rhodesians would be different than the "chums" (Englishmen) who dominated the Northern Rhodesian Game Department. Shirley supported my decision, and feeling a bit disillusioned herself, decided to give up pursuing her master's degree. She planned to move to Salisbury and take a job as a botanist in the herbarium there. I informed Director Parnell of my intention to leave should I get the opportunity to transfer to the Southern Rhodesian Game Department. While saying he regretted my decision, he agreed to not oppose the change.

My plans had to be put on hold when I was unexpectedly called up by the army in which, like all young men, I was a reservist. The call was for routine training but coincided with the Suez Crisis of 1956, for which Britain appeared likely to mobilize her empire's forces, and the Hungarian uprising of that same year, which might also have led to mobilization. I had no idea what the future held, but I obeyed orders and completed the three months' continuous training at Llewellin Barracks near Bulawayo in Southern Rhodesia. Since I had already passed through basic officer training at Plumtree, I was made platoon sergeant. The regular army instructor assigned to us was a Scot from the British army, Sergeant Derry McIntyre, who became a good friend, despite difficulties that arose later when Derry became a general and our roles placed us in opposition.

The deployment to Suez never materialized, nor did one to Hungary, but while at Llewellin Barracks I received an unexpected notification from Parnell. The Northern

Rhodesian Game Department wished me to return, and should I agree, I was to be posted to take over the Northern Province (a province nearly the size of Great Britain that was ultimately split into three provinces—Northern, Luapula, and Muchinga) as the provincial game officer! My godfather Eustace Poles would be serving under me as a senior ranger at Mpika. This was an amazing turnaround. It was also unheard of for anyone at my age, now twenty-one, to be a provincial game officer.

I left Llewellin Barracks a civilian once more, with no expectation of further active duty, though I remained enrolled as a territorial soldier now assigned to the Third Battalion Royal Rhodesia Regiment (Northern Rhodesia). I immediately contacted Eustace, who knew of the department's offer and assured me he had no problem with me being his boss. I could now see my way to doing serious wildlife work, but I had much to learn as I had no training of any sort in administration or colonial office procedures and had never led a team near the number of men I would oversee—nearly two hundred game rangers, fishery officers, tsetse fly officers, elephant control guards, game guards, and game scouts! My devotion to the game department was renewed. This new role was well beyond any I had imagined myself playing.

CHAPTER 5

TO UNIFY OR DIVIDE

Seeing Beyond the Color of One's Skin

1957

It happened when I gave an order and for the first time, my men disobeyed.

One of my courageous hunting companions, Kapombe Mukumpulwa.

I now had some serious planning to do, and fast. My girlfriend Shirley, who had moved to Salisbury, did not want to be so far apart if I was headed back to Northern Rhodesia. So, we decided to marry and that she would move with me to Kasama, the administrative center of the Northern Province of Northern Rhodesia, where I would carry out my new and daunting duties as the provincial game officer (PGO).

We were married by a magistrate with no announcements and just one close friend as witness. There was no honeymoon; the day following our marriage, I hitchhiked back to the game department's headquarters in Chilanga, declined the long wheelbase Land

Rover I was issued, and instead chose a more practical short wheelbase one, in which I drove the 500 miles of earth road to Kasama. Before joining me, Shirley had to complete her period of notice at the herbarium, after which she drove 800 miles of desolate road in a small Ford sedan I had bought a few days before our marriage.

On arriving in Kasama I was invited to a sundowner party by the PGO I was replacing, Constantine Benson, known as Con "Birdie" Benson for having published *The Checklist of Birds of Northern Rhodesia.* At the party I was introduced to Birdie's administrative secretary who was to become my secretary—"Pooch" Bathhurst-Brown, the wife of the provincial forestry officer (PFO). Upon meeting her I couldn't stop myself from blurting out, "Good God, you're old enough to be my mother!" which she wasn't, actually—and she never forgave me. Nonetheless Pooch was wonderful. She gave me all the training I would ever have on colonial office procedures, accounting, reporting, filing, and much more, as this was a very mature bureaucracy.

The day following the party I moved into what would be Shirley's and my new home: two thatched-roof rondavels adjoined by a covered passageway, just steps from my tin-roofed, two-room game department office. The handover began immediately, with Birdie smoking his pipe nonstop while fussing over all his ornithological notes and records. To most questions he quipped that Pooch would brief me, from budget details to the stationing of staff.

Birdie did take the time to elaborate on a particular staff member—a fellow by the name of Miller who he had arranged to transfer to Kasama. He referred to Miller as an "under ranger," a term I'd never heard before, stating that the unusual rank was because Miller was "Colored." I asked why the under ranger was coming to Kasama where I would have no use for him being so far from all our main game reserves. Birdie confessed that while he had told Miller he was being transferred to handle problem elephant control from headquarters, the real reason was that Miller was "a thief, a liar, lazy, and useless."

Astonished, I asked, "Well then, why don't you just fire him?" Birdie shrugged his shoulders, saying they lacked hard evidence, and it was my job to find evidence and fire this useless man. Already the shining duties I had envisioned were tarnishing—how on earth was I to deal with Miller?

Shortly after Birdie's departure, Pooch came into my office to announce Miller's arrival. I walked outside to find Miller, a pleasant-looking Colored man covered in dust, alongside his wife and two children. I introduced myself, asking if his wife would mind waiting for a moment—might she enjoy a cup of tea with Pooch while we talked? Once inside my office, I quickly shut the door and sat him down. "Miller," I asked, "Why are you being transferred here?" He explained that he was to handle all elephant control from Kasama. I quickly snuffed that vision. "Miller, that is a lie. The real reason you are being transferred to Kasama is because you are lazy, a liar, and a thief—and it's my job to catch you and fire you."

Miller's grin turned to complete dismay and shock. We sat in silence for a moment and then I said, "This isn't a good start for either of us, so why not begin with a clean slate?" Miller seemed confused, so I explained. "Miller, I want you and your family to stay with my wife and me tonight and then tomorrow drive on to the Bangweulu Swamps.[1] I want you to develop a Lechwe Control Unit down there." I showed him on the map about where I wanted him stationed. "Find a camp site," I said, "and build a camp for you and your family. I'll send you what staff I can and what money I can, although we have no plan and no budget."

Miller never let the department or me down. I knew the district commissioner[2] at Samfya helped Miller wherever he could, including providing him a boat, but where Miller "obtained" most of his building materials and more, I never asked. Miller performed wonders, building a rough house and office in no time and immediately curbing the rampant poaching of lechwe (an aquatic antelope). Poachers were ruthlessly killing them to supply a market in the Copperbelt, which lay across the Congo Pedicle that juts into Northern Rhodesia. I had heard talk about giving lechwe special attention ever since my university days, and now finally it was a reality—we had a Lechwe Control Unit (albeit only one man and his family camped under trees), and Miller, far from being lazy and useless, was to prove himself more worthy than many a game ranger.

I've mentioned coming from a family that was, like many other White Rhodesians in those early days, paternalistic in our treatment of our Black countrymen and women, who we referred to as "Africans." However, many Whites in the Northern Rhodesian Game Department could only be described as racist; their treatment of "under ranger" Miller is but one example. The men with these views unfortunately included my godfather Eustace. Eustace had revered the Karen tribesmen of Burma that he and others had raised to fight the Japanese behind their lines. He spoke often of their courage, hardiness, and loyalty. However, with Africans Eustace was very different; he had little good to say of them except for the few men he admired who worked closely with him. Later, Rhodesian politics also drew us to opposite sides, Eustace supporting the racist vision of then prime minister of Rhodesia Ian Smith and I going to great lengths to oppose it. Rather than damage our fundamental relationship and love for one another, at the time I opted to not engage with, as far as I could, Eustace's racist views.

I strived to lead by example with my staff, sharing all hardships and not expecting them to do anything I would not do. Unlike most game rangers, I never used a tent. My staff did not have tents; therefore, I refused one. I slept on the ground as my staff did.

---

[1] The Bangweulu Swamps was a turbulent area, rife with poaching due to its proximity to a forty-mile-wide spike of territory called the Congo Pedicle that projected deep into Northern Rhodesia.

[2] Each province in Northern Rhodesia was divided into several districts, each headed by a district commissioner (DC) who reported to the provincial commissioner (PC) who reported to the governor appointed by Britian. Though the game department was independently administered at the provincial level, district commissioners were generally supportive of game department efforts within their districts.

But I was still a product of my times. Like most young colonials brought up in Rhodesia, and perhaps heavily influenced by Plumtree's formative days, I believed my role was to lead by example. When I gave orders, I expected them to be followed. I did not fraternize with my staff and always slept a short distance from them by my own campfire. "They" were different and not quite like "us." That was how we had been taught to behave and, frankly, the staff too behaved as if it were the natural order of things. But I was about to experience a life lesson that changed this perspective. It happened when I gave an order and for the first time, my men disobeyed me.

On this assignment we were dealing with crop-raiding elephants in villages on the border of the Serenji and Mpika Districts. I was with two elephant control guards, Lingashi Mwamba and Kapombe Mukumpulwa. These two men, twenty years older than me, were experienced elephant hunters whom I respected. We were to hunt together many times with total trust and faith in each other. Lingashi, a powerful man whose hands and arms made his rifle seem small in comparison, carried himself on spindly legs and was always good humored, though quiet. Kapombe, a tall slim man, was also quiet, unassuming, and superbly organized. The first night we camped near one of the crop fields, listening to the elephants feeding in the dark. By dawn, the elephants—well versed in being shot at for their "crime"—had put a few miles between them and the crop field. We tracked them for hours (elephants on the move walk at close to six miles an hour whereas we, even walking fast, covered about four miles an hour). Midafternoon we caught up with the culprits when they chose to linger, feeding—approximately twenty satiated elephants in an open woodland. By then we were so far from the crop field that I decided not to shoot any because they would not associate the punishment with the "crime" of the previous night's crop-raiding. I simply sent the herd off in panic by firing one shot over their heads. Then we wearily turned back toward camp.

It wasn't long before it started to rain—a slow drizzle that turned into a downpour. As we trudged along, we startled a reedbuck ram, which darted away only to stop sixty yards off and peer at us through the rain. Tired and hungry, and knowing we had no meat back at camp, I promptly shot the ram, ordering my companions to quickly cut off the legs and a few other bits so we could get on our way.

Much to my surprise, Lingashi and Kapombe refused to obey my order, stating they would not leave any meat behind. "What do we say to the cook in camp when he asks about the rest of the meat? We left it for the hyenas because we were lazy and tired? No! We are elephant control guards in the game department, and we will carry all of it." Their message was clear, and it hit me like a brick to my head—they, like me, were proud. While they removed the animal's entrails, I cut a carrying pole using a kazembi (small axe). We slid the pole through slots cut in the ram's legs. Gutted, it weighed about 120 pounds. One of the men hoisted up one end of the pole and I, the other.

We trudged through the mud and rain, the carcass hanging between us. Periodically, Lingashi and Kapombe would swap places carrying one end of the pole. Each time they asked to stop and change, I let them change while I stayed put. Their disobedience had pricked my pride, and I was determined to carry my end of the pole all the way to our destination. To my relief, we reached camp around midnight. My shoulders and neck were numb and my head pounding; we had walked more than forty miles in nineteen hours. As I fell asleep, I knew these men had forever changed me.

During that agonizing walk, I had reflected on how the color of one's skin had created such unjustifiable division and beliefs. Yes, some people were weak, bad, dishonorable, cowardly, and cruel and others strong, good, honorable, brave, and kind, but these characteristics had nothing to do with race, color, or tribe. On that day I vowed to never let race cloud my character judgements—a great lesson, thanks not only to Miller, but to Lingashi and Kapombe, who would live on as my infallible hunting companions through many adventures.

CHAPTER 6

MY MANQUALA

Benefitting from a Bit of Magic

1957

I said nothing about my dream but cautioned, "My manquala tells me this hippo will get me."

My manquala, which is still with me to this day.

Of the Black men in the Northern Rhodesian Game Department, the majority were illiterate and only a few could speak English, so it was necessary to have knowledge of and conversational skill in their local languages. In fact, all White officers were expected to pass government language exams for fluency in one of the main African languages within two years. If an officer did not do this, he remained at his original salary level, no matter how senior he might become. In my role as provincial game officer

(PGO), I had to immediately get as fluent as I could in a local language. In my first year, not knowing where I would be stationed, I had begun learning Chinyanja, the easiest to learn of the local languages.

In Chinyanja, there is a notable word, *manquala*, meaning medicine. Many Africans had deep respect for the power of shamans, or witch doctors, and the strength of their manquala. I decided to make something that could serve as my manquala in case I needed it. I collected several items, including a lion's collar bone and strands from its mane, a vulture's beak, a dried chameleon, and a tiny crocodile jaw. I tied these items together using a shoelace (in those days, before nylon I made my laces from bushbuck hide). When asked by my staff about this oddity, I said it was my manquala, which I often consulted. Little did I realize that my manquala was about to gain great power and respect.

In the Northern Province, game department staff killed about fifty mostly man-eating lions a year,[1] but shot even more hippos, which routinely killed more people than any other dangerous game. One day I received a telegram from the district commissioner at Chinsali informing me that a hippo had killed two people. I rallied my hunting companions, Lingashi and Kapombe, and we started on our way. En route our Land Rover broke down, forcing us to spend a miserable night in the rain, cramped up inside, constantly batting at feasting mosquitoes. During that long night I had the most vivid nightmare I had ever experienced. I dreamt that I had arrived at a village by a river and borrowed a dugout canoe to follow the hippo, which had retreated into a flooded swampy area below the village. The hippo attacked and sank my canoe. As I was escaping—climbing into a tree on the flooded bank—he clamped onto my leg. I jolted awake, my heart pounding.

At dawn we got our vehicle going again and proceeded to the Chinsali boma[2] where the district commissioner assigned us a guide to the hippo-endangered village. On arrival, the scene was remarkably like my dream. The locals informed us that the hippo was lurking in a swampy area created by the overflowing river that ran past their village.

We decided to send canoes down each side of the river's flooded banks and one down the middle, with me poised in the bow ready to shoot. I stripped away everything not essential or waterproof, including my watch and notebook, and took only four spare rounds for my .470 rifle, anticipating a short, furious attack at close quarters. When Lingashi asked why I was stripping down, I said nothing about my dream but cautioned, "My manquala tells me this hippo will get me."

---

[1] The high incidence of man-eating lions I encountered in this region of Northern Rhodesia, was, I believe, attributable to the local slave trade. Many slaves died because they were forced to walk hundreds of miles from their point of capture to the east coast slave markets. It was along these routes that man-eating lions were endemic. Despite knowing of the reports by early African explorers, including David Livingstone, I did not comprehend the extent of this trade until I walked from Pweto on Lake Mweru to the southern end of Lake Tanganyika in search of elephant poacher paths into the Congo. On this trip I discovered ancient slave paths almost six feet wide and six inches deep that had been worn by hundreds of thousands of feet over centuries. These paths were still clearly visible forty-five years after the end of the slave trade.

[2] A boma was a small administrative center in each district where the District Commissioner and his staff resided.

Our flotilla glided downstream, the boatmen poling their dugouts as silently as possible, all of us studying the water ahead, waiting for the slightest ripple to indicate the hippo's presence. The angle of the sun made it impossible to see down into the water, so I focused intensely on the surface, ready to fire at the first movement. That hippo was indeed clever, however, and he let my canoe, in the mainstream, pass right over him unseen. I caught a movement through the clear water a moment before he surged up on my right side, his mouth wide open. Wedged in the front of the dugout, I could not swing my upper body around quickly enough to shoot. I could only lunge forward as the hippo's enormous jaws clamped down, crushing the canoe. His jagged teeth thankfully only went through my bush jacket, but I was wedged by the crushed canoe. I gulped for air and went under the water, clutching the hippo's ear with my right hand while holding my rifle in my left. The two of us spun around, the hippo unable to bite me as the pressure of the water held me against his neck. Then something hit me hard on the head, probably the dugout, swirling above. The force caused me to lose my hold on the hippo's ear, and I kicked away, intending to fire from the hip into him, but thankfully refraining from shooting (I realized that being under water, a shot would likely burst my rifle and kill me, not the hippo).

Although desperate for breath, I instead chose to sink to the bottom of the river, helped by my heavy rifle. The hippo surfaced, searching for me, while I stayed underwater, swimming as far away as I could and slipping into a thicket of reeds. Concealed, I stood up, gasping for air and emptying the water from my rifle's two barrels as silently as I could. I could see the hippo swimming around, searching for me. I could also hear the alarmed cries of the villagers, "Bwana wafa! Bwana wafa!" (the White boss is dead!). The villager who had been poling my dugout had climbed a tree and was perched high enough to survey it all. He could see where I was hidden, watching me carefully as I remained silent and deathly still for what seemed like hours until the hippo proceeded on down the river. Then the man in the tree shouted that I had escaped and directed Lingashi and Kapombe toward me. They quickly arrived in two dugouts over which I could drape my arms as they pulled me, badly shaken, to solid ground.

That hippo killed two more people that season until we were able to finish the job. By then the rains were over, the rivers had subsided, and the hippo had taken to lying up in a lagoon opposite another village. With the lagoon being on the opposite bank of the river, Lingashi and I crossed by lying down in a dugout, paddling slowly and silently with our hands. The hippo was about eighty yards away, completely submerged, occasionally poking his eyes and nostrils above the water. I sat hidden, sending Lingashi to the far end of the lagoon, where he cut a long branch, tied my handkerchief to it, then concealed himself in the reeds. While keeping himself hidden, he raised the branch, twitching it just enough that the flash of the handkerchief piqued the hippo's curiosity. Gradually the hippo reared his head to inspect the strange flutter of fabric, giving me the perfect shot.

That series of events—my dream-inspired prediction that the hippo would get me but I still killed *him* in the end—started rumors that my string of curiosities was indeed powerful manquala. A second incident, but a short while later, elevated my manquala's status even more.

This time, I traveled to the Luangwa Valley (a future national park) on an unannounced visit, hoping to surprise poachers. My team consisted of my driver, William Kasonde; my trusted hunting companions Lingashi and Kapombe; and game guard Joseph Musukanya. Because I suspected some of my staff could also be informants for poachers, I recruited six carriers familiar with the Bangweulu Swamps west of Mpika to accompany me. After convening in Mpika, we headed the opposite way, up the Great North Road, then dropped off in the bush on the headwaters of the Lufila River. From there we walked down the river into the North Luangwa Game Reserve, *east* of Mpika. After two days of trekking through the hills, I sent William and Lingashi back, relieving the team of excess supplies and carriers. I continued with Kapombe, Joseph, and four carriers. That evening, when I went over to their campfire to talk to them about the following day, I felt strangely uneasy, enough so that I warned the men coming on with me to be alert. They asked why, but I could only express that we might have trouble with a lion. And two days later, we did.

About midmorning, walking down the right bank of the Lufila, Joseph, who was in front, spotted a lioness. We all stopped in our tracks. I was second in line, followed by Kapombe, the carriers falling in place behind him. I had my 7mm rifle in my hands and thankfully slipped off the safety (although I had no intention of shooting). Suddenly there was a loud rasping grunt behind me, in the opposite direction to the lioness. Swinging round, I was faced by a lion in full and furious charge, barely five paces off. I fired instinctively into its chest. The lion dropped dead between Kapombe and me. Thinking that in such haste I might have shot badly and only stunned it, I quickly put the rifle against its head and fired again. That's when I learned that lions, shot through the heart as this one was, do just drop dead, unlike most antelope, elephant, and buffalo.

In both the hippo and lion incidences, the protective power of my manquala only increased its notoriety. Yet its power didn't stop there. Upon arriving in my office one Monday morning, I found a distraught Pooch, fretting over the fact that she had accidentally left our safe unlocked over the weekend. We checked for anything missing and sure enough, fifty pounds was gone. I retrieved my magical string of claws, beaks, and bones and summoned my staff. They crowded into my office where I sat with my manquala on my desk. I announced that the safe had been left unlocked and someone had stolen fifty pounds; I had consulted my manquala and knew who the thief was. As I talked, I made sure to make eye contact with every one of my men. I also assured them that I would not shame the thief in front of everyone, but if the money was not on my desk by the time I returned from lunch, the thief would die that night.

Sure enough, when I came back from lunch at 2 p.m., a sum of fifty pounds was stacked in a neat pile on my desk. Nothing more was said. From then on, I never had to lock my doors or have my camp guarded; I simply left my manquala hanging where everyone could see it. It's still with me today, in my study amongst my mementos, keeping an eye out for me.

CHAPTER 7

RISKING MY JOB TO DO MY JOB

Rethinking Efficiency and Justice

1957–1958

Forgiveness is easier to obtain than permission.

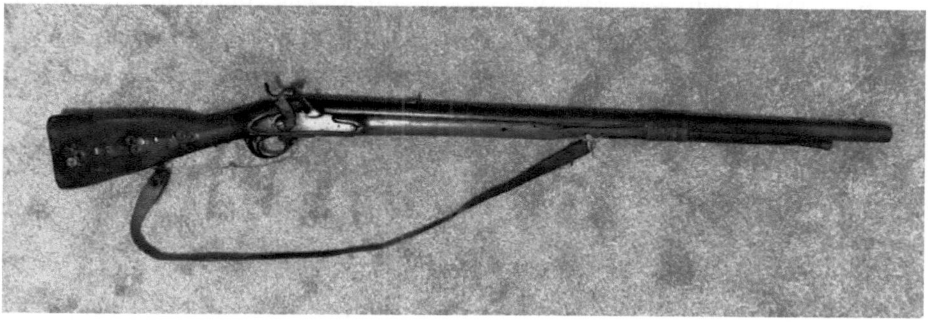

One of the muzzle loader guns popular with poachers in Northern Rhodesia in the 1950s.

Given the paucity of training I received as provincial game officer, I quickly learned that forgiveness is easier to obtain than permission. My first lesson happened when I accompanied my godfather Eustace to the public works department to get supplies. Eustace was integral to shaping my means of acquiring necessities; when a requisition form required a few different signatures he simply signed them all, using a different scrawl and pen for each (he carried six pens in his pocket specifically for this purpose). We then simply went about collecting what we needed. To get "official" permission, each signature would have required days of explaining before receiving, but via Eustace's shortcut, we had what we needed in no time. I also learned a useful lesson from Senior Game Ranger Len Vaughan, who was running the soon-to-be Kafue National Park. When the game department delivered Len a small, useless tractor instead of the larger one he'd ordered for road building, he simply drove 120 miles to the nearest town, Mazabuka, and promptly bought a heavy-duty Case tractor on a local purchase order. He knew that local purchase orders could only be used to buy items up to ten pounds but also knew that by the time he'd receive hell about it, the road would already be built—and he'd be forgiven.

This type of "efficiency" became beneficial to my management beyond supply acquisition, particularly in instances of seeking justice. The game department's handling of justice, and subsequently mine, was initially challenged one evening while conversing with my staff around the campfire. We were camped on the border with the former German territory of Tanganyika (the staff referred to it as "paGeriman"—now Tanzania), which prompted a lively discussion on German rule versus British rule. Some of the men lamented how unfair we British were compared to the Germans. Their opinions contrasted entirely with my view—Germans were so much more brutal! They then proceeded to tell me how wrong I was, that under German rule when a man did something wrong, he was punished but not his family, who were innocent. Under British rule, however, when a man did something wrong, he was rewarded, and his family punished; this was unjust and cruel.

When I said, I don't understand, they explained. When a man did wrong, the Germans whipped him with a *sjambok*[1] and it was over, the guilty man having been punished justly. The British, on the other hand, "rewarded" the culprit by putting him in jail, clothing him, feeding him well, and assigning him light work. Meanwhile his wife and children—who were innocent—were left to suffer, trying to trap for meat, protect the crops from elephants and hippo, and harvest anything left. The injustice of our British administration compared with the justice of the German administration, as viewed in their eyes, now made some sense.

The last straw was when we caught a renowned poacher from The Congo in the act. Months later, in court, the district commissioner at Mporokoso slapped a miniscule fine of about twenty pounds on the poacher. The poacher calmly pulled a pile of notes out of his pocket, paid up, and walked out. I was astounded at the laxity of the punishment. We had just had an incident on the border where poachers overwhelmed two of my game guards, nearly killing them. The poachers had handcuffed the guards to the thick hide of an elephant carcass and left them to die of thirst, which they would have had we not found them in time. I concluded that, with all we had to risk (not to mention the absurd amount of paperwork and time involved in prosecuting cases in court), I was no longer going to abide by this ridiculousness. Challenged as I was by this time with reduced staff, an increased workload, and no support from my superiors—who never visited, consulted with me, nor discussed any of the issues arising in my province—I decided to put saving wildlife at the top of my priorities, no matter the cost. Hence I began holding court in the field.

My first "case" occurred in the Luangwa Valley. We caught a man who had shot a buffalo in the game reserve. Rather than drawing out his "punishment" via a court, months later and hundreds of miles away, that would hear the man almost certainly testify that he found the buffalo already dead, I took the law into my own hands. Right there in the field I asked, "Did you shoot this buffalo?" The man answered affirmatively (had he denied it, we would have simply dug out the muzzle loader slug from the wound and

---

[1] *Sjambok* is Afrikaans for a tapered, flexible whip made of hippo hide, originally used to drive oxen.

compared it with those in his pouch, as well as pointed out his freshly fired gun and confirmed the hunt from the tracks). Pointing to the hills on the western horizon I asked, "Are those the Muchinga Hills?" Again, the man answered yes, not quite sure where I was going with my questions. Then I pointed east, toward the river, and asked, "Is that the Luangwa River over there?" Again, the man responded affirmatively.

"Are we standing between those hills and the Luangwa?"

"Yes."

"Do you know that this land between those hills and the river is a game reserve?"

"Yes."

"Did you shoot this buffalo in the game reserve?"

"Yes."

I outlined the courses of retribution: We could take the man as prisoner, handcuffing him and forcing him to walk for days with us back to court, or we could punish him right here and now and let him go. Which did he want? As my staff had conditioned me to expect, the man chose the latter.

Being mindful not to lay a hand on the man, I instructed him to pile all his belongings on the ground, including his clothes, and cover them with dry grass and branches. He then had to put his muzzle loader out of commission by jamming the barrel in the fork of a tree—permanently bending it—then placing it on the pile. Next, I gave him matches and told him to light it all on fire, which he did. He was then allowed to go.

From there on, my newly found form of justice was my modus operandi. If any man was surly, resentful, or untruthful, I would make him walk with us a couple of days before releasing him to return home naked through the bush. No one ever reported me because I stuck to what, even the criminals receiving my rulings acknowledged, was just.

Even in one case that did come to court, I must confess that I played a bit by my own rules. In this instance, I was prosecuting an elephant poacher whose case was being heard by the district commissioner in Mporokoso. The defendant spoke only Swahili, and the only people fluent enough to interpret were two game guards on my staff. I sent one of the guards as far away as I could on an assignment, and the other guard, I coached. I meticulously went over the questions I would ask as prosecutor and how he was to "interpret" the answers. No matter what the poacher said, the guard was to provide a translation that we had rehearsed. My interpreter grinned, saying he understood, and in due course we quickly obtained a guilty verdict. Yet, once more, the poacher simply paid his miniscule fine and walked off, dismissing the process as part of the cost of doing an illegal business selling tusks to coastal buyers for an enormous profit.

Game scouts, as explained in an earlier chapter, were more senior than elephant control guards and typically had more education (at least Standard VI "middle-school" or better). While some were a joy to work with, others were what we called "barrack room lawyers."

They knew everything, looked down on anyone with less education, were fond of citing regulations, and only worked the minimum number of hours. These problem scouts were also very difficult to fire. I had inherited such a man, so to get to know him better I had him accompany me on one of my foot tours. Our mission was to survey elephant damage, capture poachers, and shoot problem elephants. For days we trekked, as usual covering more than twenty miles each day. The scout seemed on his best behavior, though it didn't take long for him to break down. When he did, he point-blank refused to obey a simple instruction and spewed a litany of complaints. He wailed that I was overworking them all, had no right to make them work from dawn to dark, including Sundays, and more. I calmly took his rifle, hat, and belt, stated he was, as of right then, dismissed, and told him to find his own way back. I assured him I would pay him when he arrived at Mpika, which was well over 100 miles away.

The fellow was terrified to go on his own among all the large game in such wild bush and so slinked around behind our carriers each day, sleeping as close as possible to their fire for protection. After several days of walking, we reached the Munyamadzi River flowing through the lower reaches of the Muchinga Hills and chose to camp there before our final climb out of the valley. The next morning the game scout was nowhere to be seen. It didn't take long to find his tracks on the path ahead of us. We pondered his motive for going ahead and concluded he was attempting to reach Mpika first with the intention of reporting a very different story of his dismissal to the district commissioner. I asked my staff, "Do any of you think you could walk with me and get to Mpika today?" Lingashi volunteered and off we went, determined to walk in one day what would normally take at least two.

Lingashi and I walked fast, keeping a close eye on the scout's tracks, which at one point veered off onto a side path. Lingashi knew these paths well and said the side path went to a village some miles off. We assumed the scout was going there to borrow a bicycle. If so, it meant he would then eventually have to rejoin the path we were on and would likely arrive ahead of us. I now had to try to make sure he did not, so I told Lingashi to continue at his own pace, as I was going to run. And run I did, briskly covering the remaining thirty or so miles to Mpika.

Only in my university days, when Mike Creese challenged me to race the Comrades Marathon course, had I run further, but this time I was also lugging my rifle. I arrived at Mpika after dark, jumped into my Land Rover, and drove the several remaining miles to the boma where, famished, I enjoyed a hot meal and accommodations courtesy of Jack Fairhurst, the district commissioner. The next morning, Jack and I strolled to his office and were casually chatting on the veranda when the scout rode up on a bicycle. When the scout saw me there, he took a moment to get over his disbelief, then pedaled off. Later in the day, the scout returned, got his pay, and quietly left the department. He never filed a complaint.

CHAPTER 8

BURN THE GRASS?

My Initial Efforts in Challenging Protocol

1957–1958

What were we to do about the loss of species and their habitat, the soil exposure and soil erosion, all seemingly caused by our management, especially by our burning?

This pipe-smoking Batonka woman visited my camp near the Zambezi Valley to ask for some meat from a "problem" buffalo I had just shot. Her people were later removed from their traditional lands (this was not to make way for a national park, as in so many other cases, but for Kariba dam).

As provincial game officer (PGO), I could plan how I used my time and hence weaved the simplest of observations and research into my routine duties. This included reading and rereading every bit of ecological work I could get my hands on. I wrote my first paper, detailing my field observations on the adverse effects of the game department's policy of burning millions of acres of land every year. I argued that the policy was severely eroding the soil and adversely affecting wildlife habitat. I submitted it to *Oryx*, a peer-reviewed journal published by Cambridge University Press.

It was rejected.

Today, looking back from the perspective of old age, I recognize that my first attempt to publish what I'd observed—and its quick rejection—was the early emergence of a pattern that would repeat itself for the next sixty years. I had made observations which, though they appeared obvious enough to me, called into question established dogma and practice so radically that no one took them seriously. Nevertheless, as PGO, I could test hypotheses on the ground.

Of all the animals in the game department's care, my favorite were the elephants. No one can live in proximity to African elephants and not be in awe of their sensitivity, intelligence, family structure, sheer size, strength, and profound effect on their environment and other animals. During my university holidays spent with senior rangers Len Vaughan and my godfather, Eustace Poles, we engaged in many discussions about "elephant damage" in the Luangwa Valley. The damage was extensive and readily seen by anyone who visited. Now, with both the North Luangwa and South Luangwa Reserves under my jurisdiction, I was stirred to take a deeper look into this damage. I started by studying elephant diets, which were believed to be primarily made up of the leaves and bark of trees. I chose not to observe them feed, as elephants primarily eat at night. Instead, I chose to take stomach samples from the "problem" elephants I shot. Without a microscope or any way of storing the stomach samples, I had to do what analysis I could on the fresh samples in the field, floating the contents in a cooking pot of water while using a magnifying glass to get a closer look.

I soon discovered that toward the end of the rains and the start of crop raiding, about 60 percent of an elephant's diet was grass, not trees. I correlated this with my observations on the department's grass burning policy, which was typically enforced as early as possible during each dry season to get a green flush for the wildlife. I then matched this to the timing and location of the elephants when they raided crops. It was pathetically skeletal research, but at least something, given the vastness of the land and few resources.

In the Munyamadzi Corridor, between the North and South Luangwa Reserves, I concluded that, after the rains, most elephants fed on extensive grasslands such as those in the Chifungwe Plain. At any given time, in the 1950s, I would see 600 or more adult elephants feeding in family groups around me when I climbed a tree or anthill[1]—and that didn't take into account the calves and half-growns that were completely hidden in the tall grass. Once the game department conducted its grass burns, many of the elephants then turned largely to crop raiding.

[1] An anthill is a termite mound which, in this region of Africa, can grow to a height of twenty feet or more. The base is so large that an abundance of plants grow on it—even trees.

After my first year of burning, supported by my rudimentary data and a dose of intuition, I decided to discuss next year's elephant control with Chief Nabwalya, whose people eked out a living from meager crops, fish, and "poaching" game in the area. I emphasized to Chief Nabwalya that the game department had shot many elephants that had damaged his people's crops and grain bins. I promised him we would continue to do this in the coming year. However, one thing would change: my staff was no longer going to burn grass. I asked Chief Nabwalya if he could do the same—will his people refrain from burning grass? If he would enforce that, I assured him my staff and I would continue to shoot elephants that raided their crops. However, if his people burnt the grass, I would withdraw all my elephant control guards. On this we agreed. The following year, no one under Chief Nabwalya set the grass on fire and, as I had speculated, we had to shoot very few elephants.

At around the same time I had the good fortune of a visit from the renowned British ecologist, Frank Fraser Darling.[2] FD, as we called him, was on his third trip to Northern Rhodesia to report on the wildlife situation on behalf of my employer, the Colonial Office. Two years prior—and seven years before Rachel Carson challenged the chemical industry with her publication of *Silent Spring*[3]—FD had faced backlash from his own Ministry of Agriculture, Fisheries, and Food. This happened when FD authored *West Highland Survey: An Essay in Human Ecology*,[4] an indictment of the human factor in the destruction of nature in Scotland. The work flaunted academic convention. FD used his research budget to hire Gaelic-speaking crofters and tasked them with gathering unfiltered observations from the ground up. His findings—accusing farmers of reducing the Highland Hills to a state of desolation— scandalized his sponsors. His concluding sentence in *West Highland Survey* read, "The bald unpalatable fact is emphasized that the Highlands and Islands are largely a devastated terrain, and that any policy which ignores this fact cannot hope to achieve rehabilitation."

When Eustace handed FD over to me, following a week touring the Luangwa Reserves, Eustace warned me that FD did not bathe, and I would have to deal with it. Hence my first night in camp with FD, I had a canvas bath ready for him, but FD declined, complaining that the water was too cold! In the Bangweulu Swamps, we stopped the boat where the water was shallow enough for us all to bathe, but FD feared crocodiles, and so on it went for weeks with FD having an excuse to avoid bathing at every opportunity. Finally, in the Rukwa Valley, after we had spent two days and nights fighting an enormous fire,

---

[2] For a photo of Frank Fraser Darling see the opening of Chapter 14.

[3] *Silent Spring* by Rachel Carson was published by Houghton-Mifflin (New York) in 1962; the book documented the environmental harm caused by soldiers' use of a pesticide during World War II.

[4] *West Highland Survey: An Essay in Human Ecology* by Frank Fraser Darling was published by Oxford University Press (London) in 1955.

FD succumbed to bathing, and we all drank to that! In 1970, the Queen herself dubbed FD "Sir Frank Fraser Darling, Knight of the Most Excellent Order of the British Empire." In my opinion, "The Honorable Order of The Bath" might have been more fitting!

FD was in fact an eminent scientist whose knowledge, insights, and opinions I craved. I took my management of game reserves seriously, but had not gained the slightest guidance from my superiors other than to stop poaching, shoot problem animals, and burn the grass. What were we to do about the loss of species and their habitat, the soil exposure and soil erosion, all seemingly caused by our management, especially by our burning? Observant rangers like Eustace recognized that much was going wrong, especially in the riverine areas, but deferred to people like me—with university degrees—for answers.

Firstly, I had no faith at all in the reporting on wildlife numbers, which to me was sheer guesswork (later I concluded that the game department did exactly that, making up figures to enhance fundraising or budget allocations). I had ideas formulating on how to better save wildlife than just stop poaching (especially commercial poaching, which was far more damaging than subsistence poaching). This got me thinking—what if we gave wildlife an immediate value? In my time with FD, I proposed a concept I called "game ranching." We would not "farm" the wildlife by erecting fences and providing feed, but keep them wild by harvesting them much like wild fish in the oceans were harvested (in those days there were no saltwater fish farms; ocean fish were all "wild caught"). The game department would employ African hunters who—guided primarily by scientific research—would harvest the game. If done right, the wildlife populations would increase; in turn, the game department would earn revenue and people would have more secure employment and food.

I shared a draft of my game ranching idea with FD, asking him if he would propose it, as it would more likely get accepted if it was seen as his idea rather than mine. FD declined to "steal" my idea, however he agreed to edit my report (correcting my English as he did it) and encouraged me to go forward with it. (Two years later I was to meet, American Fulbright Scholars Archie Mossman and Ray Dasmann, who had been thinking along similar lines; we subsequently worked together to start up game ranching in Southern Rhodesia.[5])

I saw game ranching as a way to find answers to questions the game department seemed to have no interest investigating:

- What *was* the "value" of wildlife in a world in which beauty and nature were increasingly being dismissed in favor of "use" and "money" among an expanding but ever more shallow-thinking urban population?

[5] Initially, in Northern Rhodesia, I focused on game ranching to save the rapidly disappearing black lechwe in the Bangweulu Swamps. Later, I extended this idea to the conservation of elephant and buffalo in the Mweru wa Ntipa Reserve. I delve deeper into my game ranching endeavors in later chapters.

- How does one accurately determine the size, health, and productivity of a wildlife population and know what, when, and how many to harvest to maximize *sustainable* productivity?
- How does one go about harvesting humanely for productivity as opposed to sport or damage control?
- What management factors define, establish, and sustain productive habitat?
- Why did even the world's most eminent ecologist, Frank Fraser Darling, have no idea how to prevent the environmental destruction we were seeing under our own management?

These questions and more would define my scientific interest for years to come.

In regard to habitat destruction, I eagerly showed FD the damage not only in riverine areas but everywhere. In our six weeks together, we traveled hundreds of miles by Land Rover and on foot through Lavushi Manda Game Reserve (near where David Livingstone[6] had died), the Bangweulu Swamps, Mweru wa Ntipa Reserve, Sumbu Game Reserve (bordering Lake Tanganyika), and even to the Rukwa Valley in Tanganyika. I also took FD to plots the Northern Rhodesian Forestry Department had established that demonstrated the different effects of decades of burning at different times of the year. These demonstration plots, replicated in more than one province, were evidence that with fire came increased soil exposure, regardless of the time of year, and also showed how greatly the timing of the burning influenced the relative dominance of woody plants or grass plants. The deterioration was occurring over vast areas but was most visible along rivers where people—who had settled on the deeper soils close to water—had been removed.[7] This change had somehow altered the behavior of elephants and other game. I was struggling to understand exactly how, why it had a negative impact, and thus what we might do. In a report on the Luangwa Reserves to the director, dated December 4, 1957, I wrote:

> *As a result of my observations, I feel that since the elimination of humans as an ecological factor in the reserve, there has been a gradual upsetting of the balance in favour of some of the larger mammalian species which have few other enemies worthy of note. . . . I cannot help feeling that habitat rather than game is of primary importance in a reserve since the game is directly dependent upon the habitat. I feel a good reserve should be well balanced with the widest possible range of flora and fauna. This, one of the most important reserves in the country, is becoming unbalanced and will lose*

[6] David Livingstone was one of the nineteenth century's greatest explorers, as well as a physician and missionary whose efforts played a major role in ending the slave trade in East Africa.

[7] These people were relocated by district commissioners because the areas they were populating were now game reserves, some of which were intended as future national parks.

> *much of its beauty and value if measures are not taken . . . Nature cannot be left to look after herself when one of the biggest ecological factors is removed; it is no longer natural! It is absolutely necessary to take active [management] measures and to take them quickly if the situation is to be adjusted.*[8]

I talked ad nauseum with FD about habitat destruction. Finally, one memorable night in the Mweru wa Ntipa, after a good whisky, FD said to me, "Allan, why don't you shut up! You only have two options ahead of you: you will either go back to university and become a researcher, writing papers and reports to further your career, or you will go into politics." I swore blind I would do neither, and we parted the best of friends.

FD had his account of his African adventures, including his time with me, published in 1960.[9] Many years later, in the mid-1970s while visiting my godfather, Eustace, who had retired to England, we went grouse hawking in Scotland and had tea with FD. I reminded FD of his prediction about scholarship or politics and said, "You were right!" We had a good laugh when I told FD I was still fighting to save Africa's wildlife, but just as he had predicted, I was now also in Parliament.

I eventually handed FD over to my counterpart, Rolf Attwell, in the Eastern Province. On leaving, FD gave me two books that I have kept to this day. The first is titled *Kenya Diary*[10] by his friend Richard Meinertzhagen, a British solider and intelligence officer whose military accounts are now controversial. The second is FD's personal copy of Sir John Russell's *Soil Conditions and Plant Growth.*[11] Inside the cover, FD inscribed, "Remembering a happy *ulendo*[12] July–September 1957." This book was pivotal in directing my attention to soil as the key to saving wildlife and ultimately humans.

Unfortunately, FD had no more idea than I did about what we should do to better manage the reserves. Solutions would not come for many years, and to this day we struggle to implement them. At the time, unknown to me, expressing my views would once more ignite an effort to get rid of me.

---

[8] In this report I used terms like "species competing." Later, after much more work, I recognized that in nature, at the biological community level, the operating principle is not *competition* but rather *synergy* between plants and animals. It is at the species level that we "see" competition when in fact natural communities do not function at the species level.

[9] This account by FD is titled *Wild Life in an African Territory* (London: Oxford University Press, 1960).

[10] *Kenya Diary (1902–1906)* by Richard Meinertzhagen (Edinburgh, Oliver and Boyd, 1957).

[11] *Soil Conditions and Plant Growth* by Edward J. Russell (London: Longman's Green and Co, 1921).

[12] Ulendo means "journey" in Chichewa but in this context means "safari."

CHAPTER 9

TO PROTECT AND BE PROTECTED

Living with Man-Eating Lions, Evasive Crocodiles, and a Newborn

1958

Nature is neither kind nor humane.

Baby Jim in his camp bath at Kasaba Bay, Lake Tanganyika, Northern Rhodesia (1958).

My wife Shirley had adapted well to living in the bush. She was knowledgeable, could handle a gun, and she could put a meal together in minutes for anyone who showed up unannounced. She managed to do it all, even while pregnant. We were elated upon learning that our first child was due in April 1958. I had always had a deep but unspoken desire to have a son with whom I could share many years of hunting, fishing, and living amidst the big game I loved.

As Shirley's due date drew near, not wanting to risk a last-minute emergency trip, we checked her into the hospital in Kasama. The day was March 31. The following morning, while I was out shopping, a couple of people congratulated me on the birth of my son. I simply smiled and ignored them; it was April Fool's Day, and I wasn't falling for it. Once back in my office, Pooch congratulated me as well, to which I retorted, "Oh not you, too!" She convinced me to ring the government doctor, Dr. Jim Redmond, who confirmed that it was not a joke—I was the proud father of a newborn son! I sped to the hospital for my first sight of Jim. I found him lying on his side asleep in a crib, so small and tender and showing a hint of the large upper lip shared by all of us Savory men. On awaking, he clutched my finger with a hand so tiny it barely encircled my fingertip.

Once Jim was settled in at home and some form of the routine returned, I headed out on a field trip to the highly productive grasslands west of the Mweru wa Ntipa Reserve between the itigi thickets[1] and the lake fringe of dense papyrus. I was studying the grassland composition and collecting legumes alongside a newly appointed agriculture officer, who became a good friend, a keen Dutchman by the name of Bill Verboom. We were accompanied by elephant control guard Lingashi and were all crouched down, studying plants in a vast open grassland, when we were startled by the sight of a bull elephant, seemingly appearing from nowhere. The elephant must have been lying down, which was not unusual, but he was behaving strangely, constantly kicking at his stomach with his hind foot. I did not have a rifle with me that day, so I asked Lingashi for his—a .404—and had Bill and Lingashi stay back while I investigated. The elephant was facing into the wind, which was constant, so it was easy to approach him from behind. I crept to within a few yards where I could see under his stomach, attempting to figure out what was irritating him (we later found he'd been burnt by a bush fire). Unfortunately, he sensed my presence and swung around, suspicious. I stood up in the three-foot-high grass to let him see me clearly; however, he still rushed at me. His head was high and his ears out—a threatening sign that usually accompanies a bluff charge, so I threw my cap in his face, yelling "Whoa!" The elephant came to a stop, confused, hesitating for a moment before lumbering away. I then stupidly made the mistake of shouting at him to keep going. He promptly spun around and rushed at me a second time. I again presumed his charge was a bluff and, picking up my cap, stood my ground, ready to halt him. But this time, the elephant had no intention of stopping. At the last moment as I was about to shout and throw my cap, his huge thick head went down, ears folded back and trunk curled under, leaving me with no time to even raise my rifle. All I could do was fall backward and fire from my hip into his forehead as he bore down on me, my shot so close it left a burn mark around the bullet hole. The bull staggered, allowing me to roll to one side, get to my feet, and fire again, bringing him crashing down. As he fell, I was astounded to see Lingashi on the other side, rushing at the elephant, knife in hand, risking his life to come to my rescue. He'd seen me go down and thought the elephant was trampling me as it staggered.

The courage and loyalty I witnessed that day from Lingashi was like no other. From then on, I felt entirely comfortable leaving Shirley and my newborn son in a bush camp

[1] Itigi thickets are extremely dense and—in the rainy season—nearly impenetrable. They are a mixture of more than 100 species of woody shrubs growing ten to fifteen feet high.

for days without me if Lingashi was on guard. In my eyes, my family was as safe with Lingashi as with me (no small feat!).[2]

In the 1950s much of Northern Rhodesia was a wild and primitive country teeming with high numbers of wildlife. The dangers the wildlife posed were not to be taken lightly, and though hippos were renowned for killing the most humans, lions topped the charts for being man-eaters. The worst lion we encountered had killed thirty-five people before we shot him. For a reason we never understood, most of the man-eaters in our region were in their prime, unlike other regions where man-eaters were generally old or maimed and thus unable to hunt more difficult game. Shooting these lions around villages with family members present made it even more challenging; understandably, relatives were unwilling to leave the beloved victims' remains as bait (a surefire way to get the feasting lion to return).

Because of the tsetse fly, which carried a disease fatal to dogs, we could not use dogs to follow or bay man-eating lions, so we had to track them down. One lion continually evaded my game guards and me. If we waited in one village, he would kill people at another. Finally, I suggested we offer ourselves up as bait by walking along the paths between villages on a moonlit night. Although the chance of getting mauled was high, Lingashi and Kapombe agreed to try. That night we started off shortly after sundown. The idea was to talk occasionally as we walked along, as villagers did, with the person in front focused on keeping us on the path. The other two would stay alert for the first sign of an attack. I had my double-barreled .470 rifle. Lingashi and Kapombe each had a 12-bore shotgun. We had no torch (flashlight). For relief we changed positions periodically.

Our first night produced no result. The next night, Lingashi borrowed a torch from a villager to see if that would make a difference. It would at least be safer. I was in front when my companions started laughing, making futile attempts to stifle themselves. I inquired as to what was so funny. They pointed out that the torch's light beam was so pathetically weak, it barely made a difference. All we could do was laugh and carry on by moonlight. After a few more fruitless and sleepless nights, we reverted to getting a good night's sleep and tracking once more. We never did kill that lion.

Another notorious man-eating animal was the infamous crocodile. When I took over the province, I held a generally poor view of professional crocodile hunters who were making their living selling croc skins for the fashion leather trade. The department issued them licenses to hunt at night using bright headlamps, which for any other animal would be illegal. One day a croc hunter came knocking at my office door, asking for a permit that made no sense to me. The hunter, by the name of Stan Wood, wanted

[2] Lingashi's admirable reputation extended beyond me—the entire province held him in high regard; his nickname was Bashimpundu (father of twins), and they called me Bwana wa Bashimpundu (boss of Bashimpudu). I never heard a villager or headman call me by any other name nor use my official title.

to shoot in an already heavily hunted stretch of the Luapula River that others were no longer interested in. But he seemed a decent fellow, and I had no reasonable grounds for refusing his unusual request, so I granted the permit. I did so on one condition: that I join and hunt with him for a period. To this he agreed.

Over the following nights, I became a crocodile-hunting apprentice of sorts, shadowing Stan. I learned a great deal and gained a friend. The first night Stan shot three large crocs in the very same river that other professional hunters had claimed was shot out. Whereas the other hunters used a bright white light that crocs had learned to avoid by shutting their eyes or submerging themselves, Stan used a red light. The red light meant nothing to the crocs and hence they did nothing to avoid it. Stan also had a strategy for shooting crocs that stayed submerged. In his boat he carried several different lengths of metal electric conduit pipe, each taped closed on one end. Stan simply thrust one of the pipe lengths into the water, closed end first, and probed, feeling for a croc lying on the river's bottom. Once the pipe prodded a croc, Stan would position it on the middle of the croc's back, toward the head. Then he would simply fire a shot down the pipe. While a bullet could not penetrate any depth of water, using Stan's method it passed straight down the pipe and into the croc.

Even Stan, however, could not explain why some crocs were habitual man-eaters and others harmless. He had shot many crocs that had targeted people who were drawing water from rivers, fishing, or paddling canoes. From his experiences he had several curiosities; for example, he found that many of the crocs-turned-man-eaters were missing the end of their tails. In some river stretches or marshes, like the Mweru wa Ntipa, he reported that every crocodile seemed to be a man-eater. On the contrary, in the neighboring Lake Tanganyika, it seemed none of the crocodiles had interest in man as a meal.

I cannot speak for Stan's findings; however, in the crocodile-infested Sumbu Bay, part of Lake Tanganyika, I witnessed children playing in the water and fishermen wading quite far out, neither party seeming to have a fear of crocodiles. The crocodiles were simply not man-eaters. In Kasaba Bay, right next to it, where I had a permanent campsite, I frequently swam near a large crocodile and even got to "measure" him one day. That day, when I came for a swim, the lackadaisical croc flopped into the water as usual, but this time left a perfect print of his entire body in the sand. It measured precisely twenty-one feet and three inches from snout to tip of tail. Later, off the same beach, I often bathed with my baby son, Jim. When we finished bathing, that old croc would climb out and lie in the sun once more.

My camp at Kasaba Bay also had an outhouse—a long drop toilet in a hut overlooking the lake. The first time I had visitors—in this instance Shirley and FD—I advised them to make noise when approaching the outhouse so that they didn't startle the cobra that lived in the toilet hole. I knew the cobra's whereabouts during the day because it liked to sun on the toilet seat. Whenever I used the toilet he would vacate the seat and slither down a hole on the side, only to resume the "throne" when I left. Yet my continual assurances that the cobra was harmless and should be ignored fell on deaf ears—neither Shirley nor FD would have anything to do with the toilet, pleading that I kill the cobra. I was extremely reluctant to kill the cobra under these terms but finally gave in as squatting in the bush was not an option for the distinguished FD.

These seemingly random incidents, whether it be irritated elephants, man-eating lions, evasive crocodiles, or cobras simply going about their business, will forever haunt my mind for what they tell of the confounding contradictions between our behavior as individuals sharing a life with fellow creatures, and what general circumstances, policies, and prejudices often compel us all, animal and human, to do. The crusade today by people, primarily city dwellers, to never eat meat or kill animals defies nature, logic, and science. Humane is to me a human construct, not present in Nature, which is neither kind nor humane. In a biological paradigm, above the plants are the plant-eating birds, fish, and mammals; above them are the birds, fish, and mammals that eat them. Even a few plants eat insects. All living creatures die, and natural death is generally cruel—whether by parasitism, disease, debilitating accidents, or predation. In fact, dying in the jaws of a lion or crocodile is one of nature's most humane deaths for many animals. Few in nature die peacefully while sleeping. I know that when I write of killing animals it may sound horrific and cruel, however dying "naturally," that same animal would likely face an agonizingly slow demise. It's normal for an African elephant to perish of starvation at about seventy years of age when its teeth—required to grind coarse plant material—wear out. It's equally hard to bear seeing the corpses of elephant calves that have got stuck in the mud and, after days of struggle, surrendered to death. We are right to be concerned about animals; however, our concern should not be focused on the killing of them. It should rather be to ensure a humane death whenever killing is required to sustain life.

CHAPTER 10

TERMINATION

Dealing with Demands, Differences, and Departures

1958

My reputation as a "savage," quite unfitted to my position, had taken root.

Age twenty-one, shortly before I returned to Northern Rhodesia as provincial game officer (in this picture I'm on my bed in the army barracks in Southern Rhodesia, where I received the message offering me the position).

I was two years into my position as provincial game officer (PGO). Though I had my frustrations, rooted in the fact that the game department's management didn't always align with reality, I loved every moment of my job, considered not even the hardest tasks to be work, and was not concerned with my pay. I was roaming over an area of relatively wild land far larger than England, overseeing a staff of more than 200 men, gaining invaluable experience, and doing what research I could while handling administration. I never felt the need for a day off. Yet confounding undercurrents were in play. My godfather Eustace informed me that I was once more getting a bad reputation. Dereck Goodfellow, the friendly district commissioner at Kasama, told him to let me know I should not be so "aloof and arrogant" as it was doing me no good. This hurt deeply because again, I felt my shyness was the reason I appeared aloof, and I could never talk about it. Even Eustace, who knew me well, had no idea how much I struggled with shyness. Routine meetings under the chairmanship of Norman Price, the provincial commissioner and thus highest-ranking civil servant in the province, were a daunting trial for a twenty-one-year-old who looked at most eighteen. At the first of those meetings, I awkwardly held my briefcase in front of me because, in my fright, I'd wet my pants.

Norman was sensitive enough to understand my level of stress, as the policies of my department often conflicted with the policies of others as well as the priorities of the provincial administration that represented the British government through the Colonial Office. He began calling me in for private meetings during which he lent a listening ear and guidance. I in turn gained great respect for him—a respect that increased significantly when one day he called me in for a dressing down. On that day, Norman was visibly angry, shoving a report in front of me in which Jack Fairhurst, the district commissioner at Mpika, wrote that he had sent a messenger to Eustace to communicate that lions had killed two people in Chief Kamwendo's country on Christmas day and ask that Eustace do something immediately. Eustace did; he sent the messenger back with a note which read, "Good show; hope they get two more on Boxing Day."

I took full responsibility for the gaffe, as Eustace technically served under me, and I had been with him at Mpika on Christmas day. When he'd told me that he sent the messenger back with the offending note, I accepted his morbid wit. The reality was there was nothing that could be done immediately due to the rains—the site of the killing was a two days' walk in the bush. Eustace had assumed that Jack, who knew him well, would see the humor in his response (Rhodesians and Brits are notorious for joking in serious situations).

Norman appreciated me taking responsibility, and confided that he, too, knew how Eustace tended to "embellish" his rhetoric. This building of mutual trust would be more valuable to me than ever as I grew increasingly unhappy with the racism in our department and its effect on policy, wildlife, and the government. Norman would remain supportive of me to the last of my days in the game department.

One incident led to speculation about my qualifications as an administrator. Norman had agreed to discuss my game ranching proposal and asked me to present it to the deputy provincial commissioner and the Mporokoso district commissioner onsite at Bulaya Red Locust Camp in the Mweru wa Ntipa Reserve. We decided on the date and that each man would get there via his own means, bringing his own food, servants, and camp gear.

By the designated hour, the deputy provincial commissioner and district commissioner had settled in with tents, chairs, beds, drinks, cutlery, food, and servants as was befitting their rank and entitlement as senior officers. I, on the other hand, had rolled this meeting into my schedule of routine anti-poaching and elephant control duties and had walked there over many days and miles, arriving with none of the carriers, servants, tents, tables, and other supplies to which my rank entitled me. I had only my rifle, ammunition, a skimpy blanket for my bed, a knife, rice, chilies, and a few scraps of dried meat tucked in a small army haversack. My fellow senior officers were cleanly shaven, dressed in starched and ironed shirts. I was unshaven, dressed in a dusty shirt, rugby running pants, and worn tennis shoes—and no doubt smelly. While they sat in comfortable chairs, I squatted on my haunches as I usually did, without a chair. At dinner I sat on one of their boxes and placed my only personal cutlery, my sheath knife, on their crisp white tablecloth. Despite the obvious differences, I thought we had a good meeting; the evening ended round a congenial campfire. When I eventually got back to Kasama, however, Pooch warned me that my reputation as a "savage," quite unfitted to my position, had taken root.

My youth was often cited as a reason I was unqualified to be an administrator, and my irreverent responses did not help my case. A new game ranger was posted under me from Kenya. He seemed a decent fellow, albeit a bit showy with an unusually flashy mustache and silk ascot. It did not take long for the story to reach me that the new ranger enjoyed his job but objected to working under a "bloody kid." Wanting to put a stop to his grievances, I invited him to join me in dealing with a particularly wild band of problem elephants in a uniquely nasty brush-choked itigi thicket between the Mweru wa Ntipa Reserve and Lake Mweru.[1] In the dense bush (and wanting to observe my critic's behavior under pressure) I provoked five separate charges by ferocious cow elephant herds. My colleague cowered behind me each time we were charged. That evening, as we ate our meal, I declared that I didn't want to hear any more of his complaints about working under a bloody kid. That felt good, though in hindsight I realized I had in fact behaved as a kid in humiliating him.

In late 1958 my coveted professional world began to seriously unravel. Eustace had retired, as had senior rangers and mentors Len Vaughan and Norman Carr. Isolated as I was, my only connection with my department was the accounting staff at Chilanga. My two superiors—Fip Parnell, now formally director of the game department, and Major Grimwood, assistant director—were openly hostile. The world's copper price had dropped abruptly, and since the Northern Rhodesian economy depended on mining it, the government immediately ordered a drastic retrenchment of staff in many departments, including mine. Without discussion, planning, or consultation I was ordered to close specified stations and lay off most of my staff.

Today the old Northern Province is divided into three administrative provinces—Luapula, Northern, and Muchinga—but back then I was responsible for all of it. For reasons never explained, I immediately lost all my European staff, including Pooch, the backbone of my

[1] I could not at the time explain the reason for the wild and aggressive behavior of these elephants; I later came to believe it was associated with the centuries of hunting they had endured during which the main trade of the region was not only slaves (see footnote 1 in Chapter 6) but also ivory.

administration. I had to terminate many of my African field men's positions, yet there was no corresponding reduction of area nor responsibilities. I was still expected to handle everything, yet now only had two African clerks trained to do little but type letters from penciled drafts and file correspondence. I struggled to uphold my responsibilities at the various stations—now primarily retrenching game guards, elephant control guards, vermin hunters, and drivers; working out compensation for their gardens and unharvested crops; and arranging transport back to their homes. My men had to be paid, accounted for, and documented by clerks in Kasama. I would then drive hundreds of miles overnight to the next station and repeat the process. When I expressed my desire to at least discuss and plan sensibly, I received a letter back from Major Grimwood demanding that I "stop whining and obey orders."

I overdid it. While in Kasama signing papers in my office, I lost my grip on my pen and fell forward, unconscious. Dr. Redmond, the same doctor who had just months before delivered Jim, ordered me to recover in bed for a couple of days. The experience left me shaken; however, it has since enabled me to recognize signs of stress and back off before it's too late.

My final order by my superiors was to close my Kasama office and move my family and what staff remained to Mporokoso. A cursory scan of a map showed this move to be unwise. It was my farthest northerly station (reached via a 100-mile earth road from Kasama[2]) and consisted of little more than a house and a two-room office. Most communication at remote stations had to happen over slow mail—messengers on bicycles—or telegrams (we were still in the days of morse code, before radios). I was now down to a handful of staff to run the entire province, yet none of the responsibilities had been reduced. I had no secretary nor game rangers—even "under ranger" Miller was transferred to Kafue, tasked with building a tourist camp. Of my close staff, I could only retain Lingashi, Kapombe, and my driver William.

Fortunately, William (who had also been the driver for my predecessor, Birdie) had accrued an amazing amount of information from his vast network of friends and contacts. His countless acquaintances enabled him to discreetly channel information to me that I would never have picked up from our game guards (a number of whom I was almost certain were in cahoots with poachers). Crocodile-hunter Stan Wood was also an integral part of my intelligence gathering; as a croc hunter of ill repute, Stan gained information on elephant poaching in The Congo and other areas that no government official could acquire.[3] On separate terms, I had been quietly recruited by the intelligence branch of the Northern Rhodesian police to keep an ear to the ground regarding the rapidly changing

[2] Stationing me at Mporokoso made no sense. It left Mpika without a game ranger or senior staff. It was an obvious move by Director Parnell and Major Grimwood to muck me about. My suspicion was confirmed when their first move after my resignation in 1959 was to appoint a new provincial game officer and base him at Mpika, not Mporokoso.

[3] Many years later when I met Stan Wood in a retirement home near Salisbury, he showed me a file containing every permit I had issued him as well as hand-written letters from me about our intelligence-gathering activities. When I asked Stan why he had kept all of it, he replied that I was the only government official who ever treated him well.

political climate. The "Wind of Change," as British Prime Minister Harold McMillan would famously call it in 1960, had already begun to blow in Central Africa.

Once settled in Mporokoso, I hoped to get back into the rhythm of the work I enjoyed, but Major Grimwood informed me otherwise. Administrative tasks were to take priority over all fieldwork until retrenchment and reorganization were over. I tried to convince both Director Parnell and Major Grimwood that with better planning we need not sacrifice vital fieldwork, but my efforts were futile. They implied that I, too, might be retrenched, but I would rather lose my job than see all our work in the field shut down, so I again sought out "efficient" ways to do it all.

A lot of the paperwork was unnecessary, petty correspondence from bored bureaucrats in Lusaka and Chilanga making ridiculous requests such as, "How many pairs of boots are in your stores?" My two remaining clerks knew only the established protocol for dealing with incoming mail—register every letter or telegram in a ledger, file it, and place the files in my inbox for attention. I now told them to stop registering every letter or telegram and instead, simply open mail and place everything on my desk with a rock on the pile to hold it in place.

They did so, and I in turn dealt with the most urgent items. Anything important but not urgent, I placed in a second pile. I then went off to do fieldwork, often covering great distances by Land Rover or on foot from Mporokoso to Mpika. On my return, I would go through the items now piled high on my desk, tossing most of them into the wastebasket. Again, I dealt with the most urgent items and added the "important but not urgent" ones to my growing second pile. I generally never responded to anything in my second pile unless there was a second request. Amazingly, with this process I significantly reduced the amount of paperwork, kept all administration going smoothly, received no significant complaints, and accomplished much in the field.

Despite the insurmountable pressure put on me by Director Parnell and Major Grimwood, I enjoyed a certain peace of mind; I was doing what I loved in places where I wanted to be. With Shirley and baby Jim comfortably ensconced in our game department house at the boma, I would spend weeks at a time on foot tours in the wildest country. There, I was happy and superbly fit, living on a diet of mostly meat, black tea, mealie meal, bush spinaches, occasional fruits, and at times eggs and pawpaw (papaya) gifts from villagers. Walking with a heavy rifle day after day often left me too exhausted to eat in the evening and I would quickly succumb to sleep, wrapped in a blanket by my fire. At some point during the night, hunger would overtake sleep, and I would wake to eat a meal of meat ready by the fire and gulp down water, both of which I would seldom do again until the following night.

Other than waking up to eat and drink, I usually slept like a log. One evening, after a long hot day in the Luangwa Valley, we camped close to the river, and I set up my fire and bed a short way off, as usual. I had only my light blanket, which I wrapped around me tightly before falling asleep on the ground. I woke once, in the dark, to eat and drink.

In the morning, after shaking leaves and dust off my blanket, I discovered rhino tracks surrounding me. I asked the others what had happened. They looked puzzled. How could I not know? In the semi-moonlit night, a black rhino had headed down to the river and nearly stepped on me, fast asleep in his path. The rhino then apparently sniffed my body, snorted, and stamped its feet, numerous times. Having failed to get a reaction from me, it eventually walked around me and continued on its way. Lingashi and the carriers had assumed I knew and was wisely choosing to lie low. Thankfully, no one fired a shot or did anything to startle the rhino, which would have certainly made the situation dangerous.

Within a month of my last foot tour in the Luangwa Valley, I sensed I was going to be removed, one way or another. I still had not received a single visit from my superiors or other counsel, nor a formal evaluation. Twice, Director Parnell was in my province but never bothered to connect with me. I learned of his visits when Norman mentioned them, at which time I pretended to know so as not to let my department down. I had passed both written and spoken local language examinations, which should have confirmed me as a permanent officer, but my confirmation was postponed. I was informed, via a letter, that my confirmation was pending rectification of "weaknesses in my character—mainly operating out of self-interest and disloyalty to the department as well as weak administrative ability." As I could only appeal to my two superiors, who had authored the letter and were determining my fate, the writing was clearly on the wall.

Fortuitously, I had someone to air my concerns with—the legendary herpetologist Constantine John Philip Ionides, also known as CJP, Snake Man, or simply Iodine. Despite his Greek name, he was a quintessentially British eccentric. Iodine was short, thin, chain smoked and barely ate any food. During his first evening visiting me, Shirley treated him to a meal of elephant trunk, which I had cooked in a hole in the ground. Iodine ate only one thin slice of the delicacy and passed entirely on the vegetables. Amongst other things, Iodine was on the hunt for snakes that associated with human habitation. He asked if I would mind if he searched my office. To my amazement, he caught three.

Iodine confided that he never had any trouble sending live snakes, including cobras and mambas, to the snake parks in South Africa. He simply would place each one in a cardboard box, cover it with brown paper, label the box, and list the contents as "Herpetological Specimen of No Commercial Value." He would then send each snake off by regular post. Apparently, every snake got to its destination in good shape (the snakes did not require constant feeding and had enough oxygen). When I asked about customs clearance, Iodine just chuckled, saying "Good luck to them if they open a parcel!"

I told Iodine I was having problems in the department and inquired about possible game work in East Africa. He said there was a new post in the Tanganyika Game Department for an elephant control officer in Southern Province to be stationed in

Nachingwea. Would I be interested in that? I jumped at the opportunity; this might offer me the chance to seriously begin studying my most loved animal, the elephant, again. Pleased that I wanted the job, Iodine said he would recommend me to the head of the Tanganyika Game Department, Gerry Swynnerton (who I had already met when I took Frank Fraser Darling to meet him in the Rukwa Valley).

I was soon in touch with Gerry, and I appreciated his transparency. He wrote, stressing how remote Nachingwea was and how much country I would have to cover. He asked, "Would it be of concern that phones don't usually work, and you won't be able to receive mail or collect pay for months?" I wrote back, "Not at all—I'm ready for the job!"

Through December I waited anxiously for a letter or telegram from Gerry officially confirming my new job and providing the necessary logistics, but nothing arrived. This however did not alter my decision to leave; I moved forward, determined to find another position. I resigned from the Northern Rhodesian Game Department, packed up our modest Ford car, and headed with Shirley and Jim to Salisbury.

Parting with my remaining core staff, who formed a farewell parade outside my office, was highly emotional. We had been through a great deal together; their courage, loyalty, and integrity had created a bond that transcended education, wealth, age, race, tribe, or political persuasion. With tears in my eyes that I hoped they would not see, I expressed my gratitude for forever being changed by them. Such was the environment and culture of those times, however, that while we saluted each other with true and deep respect—probably even love—we did not even shake hands.

My driver William sent me letters for some years after; those ended without ever learning of his fate. William wrote that Sondashi (the elephant control guard who nearly accidentally burnt up our lorry and saved me from a lion's charge) had died in a veld fire, and my Mporokoso office was burned to the ground by the United National Independence Party (UNIP).[4] William also shared the tragic fate of Kapombe. In one of his letters, he included a newspaper cutting with the headline, "Human Skull in Court." Shortly after my departure, Kapombe was sent on his own to patrol, whereupon he was murdered and his skull for some reason produced in court. Kapombe would have known of the risk (I had made it a rule that no game guard or elephant control guard could operate alone in the most dangerous spots) but would not have disobeyed an order. Predictably, he was ambushed and murdered. Years later, in remembrance of my courageous hunting companion and friend, I named my first farm Kapombe Farm.

[4] The United National Independence Party (UNIP) was dedicated to ending colonial rule in Northern Rhodesia. Following independence, when Northern Rhodesia was renamed Zambia, UNIP governed from 1964 to 1991 under the presidency of Kenneth Kaunda. It was the sole legal party in the country between 1973 and 1990.

PART III SOUTHERN RHODESIAN GAME DEPARTMENT

CHAPTER 11

BUT THEY'RE THE EXPERTS!

The Tsetse Department's Flawed Policies

1959

All of us took our chances, being bitten often but assuring ourselves that only a small percentage of the flies carried the parasite fatal to humans.

Shirley, Jim, and I having a swim at Dotts Drift in Southern Rhodesia. Little did I imagine Jim would one day grow to be six foot four and tower over me.

On arrival in Salisbury, Shirley, Jim, and I stayed with my father until I could secure a job. The Southern Rhodesian Game Department had no openings at the time; however, the director of the tsetse department, John Ford, found a spot for me in his department, knowing it would be short term until something opened in the game department.

Shortly after accepting Ford's offer, I received a letter from my former secretary, Pooch. Pooch wrote that she had stopped for tea at the Mbeya Hotel in Tanganyika and encountered a letter addressed to me, pinned to the hotel's message board (hotels provided message boards in those days for people living in the area who had no other way to send or receive mail). It was the long-awaited confirmation from Gerry that I had the job he had offered at Nachingwea! Apparently, a clerk in the Dar es Salaam Game Department had addressed it to me at Mbeya, Tanganyika and not Mporokoso, Northern Rhodesia. Having already committed to Southern Rhodesia's Tsetse Department, I now had to turn Gerry down. With regret for what might have been, I wrote back, letting him know what had happened and thanking him wholeheartedly.

The tsetse department built field camps using prefabricated metal huts, and I quickly made a start on erecting a new home for my family in the Sabi Valley at an old wagon ford known as Dotts Drift. I'd found a nice site on the Sabi River (now the Save River) where I terraced a steep bank, created an open-air kitchen, then built a dining hut and bedroom hut under the shade of a fine grove of trees. Our new home was thus long and narrow, with the kitchen at one end and the toilet at the other. The toilet was a seat over a pit, surrounded on three sides by a reed mat. The fourth side opened onto a view across the river, encouraging meditation. As soon as camp was finished, Shirley, cradling Jim, drove the 200 miles from Salisbury to join me.

We barely settled in before we were instructed to move further south to a site on the Lundi River (now Runde River) opposite the Chilojo Cliffs. I did round two of building us a home. Three months later, during the hottest time of the year, I was transferred to the lower Zambezi Valley. There I did round three of constructing a similar camp of metal huts. The third time's a charm, right? Or so I wanted Shirley, who was incredibly patient with the continual changes, to believe.

My position title was entomologist, and I reported to Dr. Fido Phelps, a senior entomologist and a friend from university days. My only neighbor, Tsetse Fly Officer Alan Simmons and his family, lived in an old mission house at Rekomitjie. As there were no other people living in the area below Chirundu, I was free to wander the whole river and valley. The tsetse department started using radios to communicate with the many field stations and installed one at Rekomitjie, ordering us all to report via radio daily at 8 a.m. To remain in camp and begin work well after eight made no sense, as it was smarter to take advantage of the cool mornings before the stifling heat set in. I elected instead to start work shortly after 5 a.m. and was often miles from camp by the 8 a.m. radio check-in. My colleagues at headquarters typically left their desks by 4 p.m., and I often did not return to camp until sundown. I simply did not communicate most days, later reporting radio "problems."

The tsetse department, officially the Department of Tsetse and Trypanosomiasis Control, oversaw just that—controlling the rather nondescript, biting tsetse fly. The tsetse fly carries parasites of the genus *trypanosoma*, which infect domestic animals, especially dogs, horses, and cattle, usually killing them. Two subspecies, *Trypanosoma brucei gambiense and T. b. rhodesiense*, also cause a fatal sleeping sickness in humans. All of us took our chances, being bitten often but assuring ourselves that only a small percentage of the flies carried the parasite fatal to humans.

It seemed the worst characteristics of bureaucratic pseudoscientific thinking drove the tsetse department's policies. Southern Rhodesia's main measure to eradicate the fly was to create tsetse-free areas for livestock by ordering the eradication of *all* wild animals on whose blood the tsetse flies thrived. In other words, kill off hundreds of thousands of animals so that flies—without blood for food—would die.

Where I had previously hunted problem elephants and always shot the minimum numbers, now for the first and last time under department orders, I shot an entire herd of fifteen elephants. Once the herd matriarch was dead, the remaining cows and calves milled around in a terrified, trumpeting huddle, meeting their deaths one by one. It was sickening, and I never did it again (from then on, I have always quietly scorned the men who brag about how many elephants they've shot in government culling operations).

Entomologists and veterinarians justified the horrific operation after noting that the fly belts (the areas containing tsetse flies) had receded following a great rinderpest plague[1] that had swept down Africa, killing more than 80 percent of the cattle and countless wildlife. They chose to ignore evidence that the fly was spreading back through all the areas where the fly had disappeared—or to query why this was happening. The eradication policy resulted in more than a recorded 700,000 head of game being killed by government hunters. Thankfully, I was now able to play a role in ending it.

Director John Ford was not in favor of the obviously failing policy and it took little effort on my part to gather data to support him having it closed. In turn, with Ford's and Dr. Phelps' support, I began to design a research project to determine the main factor in the expansion of tsetse fly populations. Was it an increase in their food (the blood supply of animals)? Or was it an increase in breeding success due to habitat change that produced an increase in breeding sites? I first pointed out that in "completely shot out" areas, there were still plenty of flies. To prove it, I trapped flies, with little difficulty, twenty miles beyond the game-fenced shooting areas intended to block the fly. I also found tsetse flies in areas where, mandated by the department, herbicide had been used to kill off the trees and shrubs, with the thinking that tsetse flies wouldn't thrive in open grassland habitats.

My field observations indicated that some of the department's policies might actually be increasing the fly population by creating more suitable breeding sites for this slow-breeding insect. My thesis for this was rather simple. Tsetse flies are unique in that each female may live as long as three months but produces very few young (only five to ten). These young develop into larvae inside the female—one at a time—and are then buried, generally in sand, where they pupate into adults a month later. Surviving this long process

[1] Rinderpest, also known as cattle plague, was a highly contagious and deadly viral disease that affected cloven-hoof animals. It was officially declared eradicated in 2011, making it the second disease (after smallpox) to be eliminated from the planet.

is critical to a tsetse fly's breeding success. Ironically, due to the constant burning on the land to ease visibility and tracking for hunters, the land was suffering from a decrease of fibrous-rooted grasses and a subsequent increase in bare ground, soil erosion, and tap-rooted woody plants. This appeared to offer the tsetse fly an abundance of gullies and sand deposits for depositing their larvae, thus increasing their breeding sites.

Though we ended the tragically wasteful policy of annihilating wild animals to control tsetse flies, attempts to control the tsetse fly with herbicides and insecticides nevertheless prevailed. Eventually, years later, more sophisticated techniques for inhibiting fly reproduction were implemented, specifically sterilizing male flies.

The Veterinary Department's polices for controlling foot-and-mouth disease[2] were comparable in absurdity to those of the tsetse department's for controlling tsetse fly. Veterinarians claimed foot-and-mouth disease was mainly spread by buffalo. To contain it, the department mandated that hundreds of miles of high tensile steel game fences be built. Hundreds of thousands of animals of many species subsequently died in Southern Rhodesia and Botswana when the fences cut them off from their water sources.

Again, I questioned the policies of "experts" and studied their own data from a broader point of view. The senior veterinary researchers provided me with a report on every outbreak of the disease ever recorded in Southern Rhodesia and I noted the location of each one with a pin on a map. The distribution of the pins on the map showed no correlation with buffalo populations. What it did show, however, was a strong correlation with the areas experiencing the greatest environmental deterioration.

When I shared this evidence with my colleagues, they triumphantly pointed to the lone pin correlating with a significant buffalo population. The pin was at the headquarters of Wankie National Park (now Hwange National Park), where foot-and-mouth disease had broken out in a herd of dairy cows kept by the staff. However, when we investigated that outbreak, it appeared to have been spread by a veterinary officer traveling from a foot-and-mouth afflicted area in Northern Rhodesia. He had visited his girlfriend at the park headquarters a weekend before the outbreak. The evidence simply was not there to support the belief that the movement of wild animals, especially buffalo, was the main cause of foot-and-mouth disease.

I also pointed out that were I to design an experiment to prove whether wildlife or humans were the main spreaders of the disease, I could do no better than the cordons the Veterinary Department established around every outbreak to curtail the spread. Every road, well-used bush track, or path leaving or entering the area had veterinary staff posted. In these places, the staff had people dip their shoes in a chemical bath, cross their vehicles and bicycles through a chemical dip, and spray all tires. Between these human

[2] Foot-and-mouth disease (FMD) is a contagious viral disease primarily affecting cloven-hoofed animals like cattle, sheep, and goats. It is characterized by fever and blisters in the mouth and on the feet.

paths and roads there were miles of open country. Here, thousands of wild animals were passing in and out of the cordoned area—not to mention birds and dung beetles flying overhead—but it was believed that just by disinfecting humans and our vehicles, the disease would be stopped (it generally was to the extent the policy is still in place half a century later).

Slowly, but increasingly, I was realizing that policies based on expert advice, when those experts were narrowly trained specialists, could be linked to most of the wildlife and environmental degradation I was trying to understand. Today I see this clearly, however sixty years ago few saw anything wrong with basing policies on the advice of narrowly trained experts, and neither did the policy makers. Rather, it was encouraged and still is today.

A valuable lesson was to emerge from the flawed tsetse fly and foot-and-mouth disease policies: management decisions that destroyed the habitat of wild animals were far more dangerous and caused more lasting damage than game culling, hunting, or poaching. Yesterday's eradication areas are today some of our country's best game areas. Where we did not seriously destroy the habitat—despite government officials shooting thousands of animals in eradication efforts—most populations quickly recovered.

CHAPTER 12

OF SPOTS, SPIKES, AND (SNAP)SHOTS

Learning from a Few Leopards, a Couple of Rhinos, and a Porcupine

1959–1962

I shot a buffalo, cut off one of its legs, and sent William back to the village to bargain for the porcupine.

This rhino got so close to my camera that only the tip of her nose filled my final frame.

It might seem that in my game department days so far there wasn't a creature I hadn't experienced some sort of "wild" encounter with—whether it be wrestling hippos, changing the minds of charging bull elephants, staring down black mambas, or tempting the appetites of lions. However, there are a few more animals worth noting that made memorable impressions on me during that time, marking my days as I transitioned into my work with the Southern Rhodesian Game Department.

In my time as provincial game officer (PGO) in Northern Rhodesia, I had experienced the extreme range of behaviors one could expect of leopards. I only had to shoot a few, but one of them gave us great problems. For some reason, despite an abundance of game, this leopard preferred to eat humans. I had found young, inexperienced leopards easy to deal with, but this one defeated all our efforts, becoming ever more cunning with each of our attempts to hunt it down. In an unusual turn of events, the leopard eventually did the job for us. A man was returning home from fishing, alone, carrying his fishing spear over his shoulder. The leopard, waiting in the dense foliage of a tree, only saw another opportunity for "lunch" and leapt onto the man's back, impaling itself neatly through the heart.

While studying elephant damage to trees I (fortunately) had a milder experience with a leopard in the Zambezi Valley. I was moving from tree to tree across the flood plain near the mouth of the Rekomitjie River, accompanied by an illiterate but courageous game scout named Scarface[1] who carried my rifle as I was preoccupied with clipboard and pencil, jotting down my observations. Suddenly Scarface started making rather peculiar noises behind me. I turned to find him frozen in his tracks, his eyes peering up in dread just above my head. I looked up from my clipboard and found myself face-to-face with a large leopard crouched in the fork of the tree I was inspecting. Our eyes locked—I held my breath, afraid to even blink. We were barely five feet apart. The leopard and I engaged in a sort of stare-down competition for what seemed like hours before the leopard finally sprang away.

Of all the animals in the bushveld, there were simply no rules on what sort of behavior to expect from a leopard. Once, driving close to what is now Chizarira National Park, I saw vultures convening in the trees and, out of curiosity, stopped to figure out what had caught their attention. Sure enough, a dead buffalo lay on the ground some way off the bush track, its predator nowhere in sight. I began to walk over for a closer look, but on second thought returned to my Land Rover to retrieve and load my 7mm rifle. Halfway back to the buffalo, I heard the most awful noise and turned to see a leopard sprinting toward me from at least seventy yards away—a very long distance for any animal to charge from. A fast-moving leopard is difficult to stop with a light rifle so I waited until I could not miss and fired at his chest. The leopard, just paces from me, broke his stride and immediately began spinning around, snarling and biting his side where the bullet had come out. It curled into a tumbling mass of fury, tearing at itself at my feet until it died. My bullet had hit it in the chest, missed its heart, and come out behind its ribs, which caused the leopard to maim itself—rather than me—before it died.

For every animal that attempted to kill me, there were also some that became my pets: pookies (a tiny lemur), hedgehogs, steenbok (a small antelope), owls, ostriches, elephants, mongooses, tortoises, monkeys, a shrike (passerine bird), and hawks, to name a few. When I originally settled in as PGO in Northern Rhodesia, I took up hawking again and, with my godfather Eustace's help, trapped and trained an adult Ayres's hawk-eagle. But I soon had to give her up because of the sheer magnitude of my work. With great sadness, I released her, wishing her well.

---

[1] Scarface, whose full name I sadly do not remember, earned his nickname when a cow kicked him in the face as a child, leaving a permanent scar.

I would take up hawking again after leaving the Southern Rhodesian Game Department, when I was given a black sparrowhawk tiercel (male) I named Chaka. We became inseparable and had many memorable hunts, including Chaka being the first trained black sparrowhawk to take an adult guineafowl (a bird much bigger than he) in southern Africa. Unfortunately, while under the care of a friend one day, Chaka, tied to his perch, was killed by what appeared to be a dog or cat. I vowed to never again place a hawk in someone else's care; this combined with the fact that I was often away meant I would also never again fly my own birds, though I forever remained a falconer at heart.

Shortly before leaving the Northern Rhodesian Game Department, I acquired another "pet"—a baby porcupine that I first encountered in a village I was driving by. A group of youngsters were holding the tiny porcupine, which wasn't more than a fleshy puddle of skin and a few straggly broken quills. I asked if they would sell it to me, but they refused, saying they were going to eat it once it had grown. Later that day, needing meat, I shot a buffalo, cut off one of its legs, and sent my driver William back to the village to bargain for the porcupine. The youngsters agreed and Chinungi (meaning porcupine in Bemba) became our newfound pet. My son Jim, now seven months old, after a couple of pricks from Chinungi, quickly learned to handle him gently. Chinungi and Jim became inseparable.

Wherever we relocated, Chinungi typically dug his own hole outside—a large, deep, cool place where he would spend most of his daylight hours sleeping. Watching him dig these holes, I realized that some of the holes I'd attributed to antbears (aardvarks)—that were often used by warthogs and other animals—were actually made by porcupines. In the Zambezi Valley however, at the start of my stint with the tsetse department, Chinungi immediately made his home under our bed. For whatever reason, Chinungi never did dig his own hole, instead finding safety sleeping under our bed until evening. Then, being nocturnal, his day would start; he'd waddle off into the bush, exploring.

As Chinungi grew, he developed lovely long ornamental quills but also thick, sharp defensive quills, and he started shedding both around camp. Not wanting Jim—who was beginning to walk—to get his eyes poked out, Shirley and I made a habit of picking up every quill we saw and jabbing it into the nearest tree, out of Jim's reach. Visitors, whom Chinungi greeted like a friendly dog, often asked if porcupines really do shoot their quills when angered. I simply pointed to the quills, now in the form of tree spikes all over our camp, to keep the myth alive.

Chinungi was supremely gentle and would roll his defensive quills to one side to expose his stomach, which he loved to have scratched. Only once did he jab one of his quills into my arm. It was my fault. I had become impatient with him as he was taking his sweet time getting into my Land Rover (Chinungi often rode with me when I went to town). So, I picked him up and oh, what a mistake! Chinungi jerked his skin and sent a quill deep into my arm. I quickly pulled it out, but it left a painful bloody hole. How porcupines "eject" their quills—and the speed with which they do so—mystifies me. After that, I built him a wooden ramp to make it easier for him to get in the Land Rover.

I never knew how far Chinungi "explored" until one early morning I had to drive into Salisbury well before dawn. I met Chinungi six miles from home, happily trotting down the bush road. He was pleased to see me, and I stopped to scratch his tummy and talk to him. Then, of course, he wanted to get into the Land Rover with me, but I didn't have time to spare so I had to leave him on the track. When I came home two days later, Chinungi was fast asleep under our bed.

Too often pets come to a sad end, and so it was with Chinungi when I at last joined the Southern Rhodesian Game Department and was transferred to Salisbury. I had no option but to take him to a nearby national park, where I asked the rangers to look after him. Chinungi, ever exploring, wandered onto a nearby farm one night, smelled pumpkins in a store, ate a hole through the door, and feasted. The irate farmer promptly shot him.

My pets were a relief from the hunt-and-kill tasks of the game department, though not all encounters necessitated shooting. In fact, in some cases, "shots" came from my camera, not my gun. In both the Luangwa Valley in Northern Rhodesia and the Zambezi Valley of Southern Rhodesia, we had a great many black rhino. Late one afternoon I made camp by the Chewore—a tributary to the Zambezi River in Southern Rhodesia—and set to work on analyzing the stomach contents of two buffalo I had shot. I had parked my Land Rover near the carcasses to fend the hyenas off and, after snapping photos of some of the vegetation, set my camera (a Zeiss Contarex) on a tripod nearby. A few paces from both vehicle and camera I prepared to bathe, pouring cold water into my portable canvas bath. I then stripped down, settled into my shallow bath, and soaped up. As I was scrubbing, a rhino bull cautiously approached the dead buffalo. It did not startle as I eased out of my bath, dried my hands, and slowly moved over to my camera. Stark naked, I took a picture of the rhino smelling one of the buffalo bodies. At the click of the shutter, the rhino looked up and advanced toward me. I took another shot and then another; at each click the rhino came a little closer. When I finally ran out of film, its horn was but nine feet away, according to the camera lens. We now simply looked at one another for what seemed like ages. I began talking to him, ever so softly. Eventually, the rhino tired of my blathering and ambled away.

A couple of months later I had a similar experience with a lone cow rhino. I spotted her in an open grassland, hunkered down, and took a picture. She, too, heard the click of the camera and moved toward me, inquisitive. Again, I sat still and let the curious creature do the moving. The rhino got so close that the tip of her nose filled my final frame. Feeling the rhino was at peace, I started to talk to her softly, simultaneously pulling out my pipe and tobacco pouch, assuring the rhino not to worry—I only intended to have a smoke. I filled my pipe, continually soothing the rhino with soft-spoken words, then lit my pipe. That was nearly my undoing. My Ronson lighter made a metallic click. The noise startled the rhino and she became agitated and dangerous; I realized she could flatten me in a heartbeat. I sat, frightened and quiet, barely taking a breath. Thankfully, the rhino slowly calmed down again. We hung out with each other for a while longer before she turned and sauntered away.

CHAPTER 13

FALLING FROM THE SKY

Controlling Quelea Finches and Parachutes

1960

When lit, the subsequent explosion blasted millions of birds in their roosting sites.

Africa's first international parachuting competition in Ndola, Northern Rhodesia (1963). I am third from left in bottom row. Bob Jones is first on left in bottom row, and Jack Masterson is fourth from left in second row (with his hand on his hip).

At the end of 1959, three new professional posts opened in the fledgling Southern Rhodesian Game Department headed by Director Archie Fraser. One position was for fisheries research, which Frank Junor accepted. Another, that of game research officer, went to Graham Child, and I became the "one-man band" quelea finch research and control officer. I did not relish this, having hoped for the game research post, but at least I was back in the game department.

I was stationed in Salisbury, where Shirley and I bought a half-built home on five acres of land near the city, and I began learning what I could about quelea finches. Few people outside of Africa know or care much about these birds; they rank high among wild creatures and plants born to exploit habitats created for them by humans. They feed primarily on grass seeds but find vast areas of grain crops irresistible and subsequently do extensive damage. They've often been called "Africa's feathered locust"—their flocks number in the millions.

South Africans already had an efficient quelea research and control unit under Dr. Dirk Lourens, who had been at Natal University with me. I arranged to spend a couple of months in Pretoria, South Africa shadowing Dr. Lourens, learning about the unit's survey techniques and elimination methods. They primarily killed the birds by aerial spraying the roosts at night with parathion. One fine droplet of this highly toxic avicide apparently was enough to end a bird's life, not to mention any surrounding small mammals. This then endangered anything feeding on the dead birds, including the people who gathered them for food. Having millions of birds die slow deaths from poison had no appeal for me so I decided parathion was not something we should use in Southern Rhodesia (since its development in the 1940s, it has been banned in many countries).

The only other technique the unit had tried prior to parathion being developed was the use of sticks of ammon gelignite, a gelatinous cousin of dynamite. They first filled dozens of forty-five-gallon drums with a mix of petrol and diesel, then distributed the drums ten paces apart throughout the roosting colony while the birds were absent during the day. They placed gelignite sticks under and on top of each drum and then connected the drums with a detonating cord called Cordtex. When ignited after dark, the sticks' subsequent explosion blasted millions of birds in their roosting sites. This seemed less destructive to me than parathion, so Dr. Lourens kindly arranged for me to do a hasty training in explosives with South African army engineers. The engineers taught me how to handle ammon gelignite and to ensure the drums—when blasted—did not simply splatter fuel but exploded into flames, the blast covering the widest possible area of land rather than going high. After having me do a few detonations on my own—testing my knowledge—they issued me a blasting license, acceptable in Southern Rhodesia, and I returned home.

Quelea flocks gather into one massive nesting or sleeping colony at night. In the following months I located their roosting sites, most prominent in the dry southern parts of Southern Rhodesia. Using a compass, I plotted their evening flight paths from different positions and noted where the lines crossed on a map. When I located a site and determined its size, I would then calculate how many drums and explosives I needed. With the help of local village volunteers (who were anxious to kill the birds as well as feast on them), we positioned the drums on end, spacing them ten paces apart, often through dense reed beds and thorny bush.

After the safe but back-breaking work of getting the drums into position, I would clear the site of villagers and spend a day laying out the gelignite charges. With the help of a volunteer game scout, I then fastened Cordtex to the sticks of ammon gelignite. Periodically I would fasten a bottle of petrol to the Cordtex, about five feet off the ground, to ensure equal ignition of the fuel. As the day heated up, so too would each drum. At a certain temperature, the drums would suddenly expand with a thump, the flat ends bending outward, sending my heart thumping too! (To prevent myself getting blown up, I had dug a hole under each drum, placed the stick of ammon gelignite inside the hole, then covered it with soft earth. This way, when the drum expanded, it hit earth not the explosive.)

I would set this all off via a battery and detonator connected to a long wire from where I would lay as flat as I could on the ground (occasionally, when short on detonators, I set off everything by simply shooting one stick of ammon gelignite with my silenced .22 rifle). Once all was laid and the trap set, we would wait until the birds were settled for the night, ignite the site, and then boom! They'd dramatically meet an instant death, the massive blast lighting up the skies and rattling windows in towns ten miles away.

The following day, since ammon gelignite, unlike parathion, didn't poison the birds and the risk of eating them was minimal, villagers would gather from far and wide with sacks to collect the millions of dead birds. People picked the ground so clean that I started weighing every sack to determine a reasonable count of how many birds we killed. This counting process, albeit rudimentary, was far more accurate than the numbers approximated from sampling density of dead birds on the ground using the spraying method promoted in South Africa—not to mention it seemed common practice for officials to exaggerate numbers to gain funding and expand their agencies.

Of critical importance, I always did a final walk of the area before detonation. During one of those walks, in the Sabi Valley close to my former tsetse department camp at Dotts Drift, I found fresh rhino spoor indicating a rhino had gone right into the blast area. Over the next hour it was no small feat flushing that rhino out of the site without being charged or him causing damage to the setup, all while flocks of queleas came flying in.

I routinely invited our game rangers as well as our director to visit, learn about, and participate in my quelea control operation, but no one ever did. After I left the department, my successors unfortunately reverted to aerial spraying using the toxic parathion.

Creating massive explosions was apparently not enough excitement for me because that same year I joined the recently formed Central African Parachuting Club operating from Mt. Hampden airfield, a few miles outside of Salisbury. Two of the main characters forming the club were Jock Hutton and Bob Jones, World War II veterans who had parachuted into France and the Netherlands on D–Day. Before long four of us—Jock, Bob, Jack Masterson (another young adventurer like me), and I—were doing almost all the free fall displays at air shows around the country.

These were early days of free fall parachuting; we had neither training equipment nor coaches. We taught ourselves by first reading about what the French and others were doing, then trying it. We used old war surplus chutes that had neither the sleeve that mitigates the opening shock (now standard) nor the blank gore that enables steering the descent. Needless to say, we experienced many crash landings. Our only training before our first free fall was jumping off a forty-five-gallon fuel drum, where we would practice hitting the ground feetfirst, then rolling to soften the landing. On one of my first jumps a strong wind blew me into a railyard and I crashed into a truck (railway car) with no hope of any roll at all. Thankfully we wore steel helmets, also relics from the war, and my only injury was a dented helmet.

One day I forgot the vow I made while in the Northern Rhodesian Game Department, after following the lion into the reeds with an empty rifle: check, check, and recheck everything! Jock had helped me kit up, both of us chatting away. As usual I dropped into the sky, free falling for about a minute, quickly reaching terminal velocity (120 mph) and then—at 2,000 feet from the ground—pulling the rip cord. Because of our old-style chutes, we had to ball up tight and absorb the opening shock with our shoulders, otherwise risking the jolt of our heavy steel helmets breaking our necks. I balled up really tight, my arms wrapped around my knees. All was fine until—as I relaxed to drift on down, my chute wide above me—I felt myself slipping out of the chute's harness. Distracted when kitting up, neither Jock nor I had double-checked the quick release box holding me by all four straps to the parachute. We had failed to twist and lock it! When I bunched up, I released all four straps and was no longer attached to the main chute. Thankfully I was just able to catch hold of the harness, to which I clung desperately all the way to the ground. Amazingly, considering our rather risky learning as we went, the club only had one fatality, and that occurred after the army had built a more sophisticated para-training facility that we were allowed to use.

In 1963, we held Africa's first international parachuting competition at Ndola in Northern Rhodesia. We couldn't have chosen a worse day—the winds were dreadfully high and constantly changing. With a crowd of spectators already gathered—and after discussion amongst the officials—it was decided to go ahead with one competition and see how it went. Our team volunteered to go first.

Our team was jumping from two planes, with me being the first to drop. All went well until, on releasing my chute (which now had a sleeve to ensure a more gradual opening), it failed to open. I quickly pulled the rip cord for my reserve chute but, in coming out of the pack on my chest amidst a tumble, it caught on my boots. I managed to get it free but the chute formed more of a bundle than a wing above me. Now I had two failed chutes, though they were at least pulling me upright and I was no longer tumbling. The reserve chute seemed my best chance at survival so I pulled its rigging lines down as far as I could and let go; on the third try—to my relief—it fully opened, breaking my fall a moment before I hit the ground.

I was in a bit of a dazed state when the ambulance reached me, and Bob—who was clinging to it—rushed to my side. The hefty, ex-paratrooper-turned-brick-layer that he was, Bob almost suffocated me with the force of his hug, so elated he was that I was still alive, as from his viewing position, my chute had opened too late.

CHAPTER 14

OF PRIORITIES AND PERMITS

My Role as Game Research Officer and the Start of Game Ranching

1960–1962

I was soon to learn that my game department—the agency whose main purpose was the protection and preservation of wildlife—had no enthusiasm for protecting and preserving wildlife, nor serving the public.

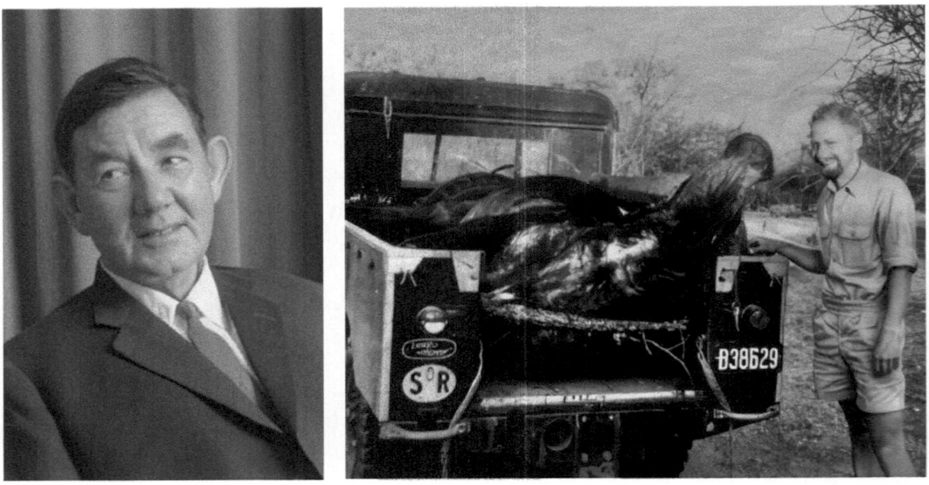

Two people who were significant in supporting my quest to understand the causes of habitat destruction: (*left*) renowned British ecologist, Frank Fraser Darling (or FD), who became a friend and mentor; and (*right*) Dr. Archie Mossman on Doddieburn Ranch with his first buffalo kill at the start of our game ranching collaboration. *(Photo of Frank Fraser Darling used with permission; © BBC Archive.)*

In May 1960, I represented Southern Rhodesia at an international conference on quelea control in Bamako, Mali. Upon my return, knowing it would be easier to get forgiveness than permission, I deliberately missed my connecting flight in Dakar, Senegal. Instead, I spent a week visiting the Richard Toll Research Station outside Dakar as a guest of Peter Ward.

Peter was a researcher from Oxford University who I'd met in Bamako. At one point in his office, we had a conversation that I'll never forget. Peter was explaining his research program on the birds of Senegal when he suddenly stopped and said, "Allan, you seem bored."

My mind had indeed wandered, though I was a bit caught off guard being called out for it. My attention was on what I was seeing outside. "Come to your window, Peter, and explain to me what you see out there." Peter joined me in looking out at a vast area of bare, dry soil and blowing dust, littered with goats and people chopping branches off the few scattered trees to feed the nearly starving animals. The ground was as lifeless as a clay tennis court. It was a glimpse of desertification in action.[1] "By the time you publish your papers," I said, "the habitat for most of the birds you're researching will be gone—as will the birds." I suggested that we needed to focus on reversing the destruction of wildlife habitat, not making careers for ourselves publishing papers on disappearing wildlife. Peter didn't disagree but couldn't see any other option for himself than to continue his work. We parted friends.

When I returned to Salisbury, I was thrilled to be met at the airport by Shirley with our newborn daughter, Megan, cradled in her arms. I had thought I would be back well in time for the birth of our second child, but Megan had other plans. Having only dreamed of having a son, I could not get over how immediately I bonded with our daughter—and how wonderful it was to have her in my life.

At the end of 1960, Graham Child left his post as our game research officer to pursue an academic career. Director Archie Fraser asked me to take on the game research job as well as mine and I eagerly accepted. I now finally had the job I wanted—overseeing game research and management for the country! I chose to focus on two main issues that Graham had not believed were priorities:

1. to learn what we could do to stop the terrible environmental destruction occurring in areas set aside as game reserves and national parks; and
2. to see if we could increase the value of wildlife outside the parks and reserves by introducing game ranching so that landowners would willingly get rid of their cattle, which all scientists including me believed were causing much of Africa's land degradation.

At that time the department was contemplating a change in game laws. As it currently stood, no one owned the game. Government did not, so that they could not be sued if the game damaged someone's property. The landowners did not, because they were not trusted (it was presumed they would shoot out all the game in favor of running cattle). Now the government was considering passing ownership to the owner of the land on which the wild animals were at any one time, apart from a few rare species that were termed "royal game."[2]

---

[1] Desertification is degradation so severe that the result is a man-made desert. The process begins with the loss of biodiversity and soil cover, and it occurs in the seasonal rainfall environments where humidity is poorly distributed throughout the year, as it is on almost two-thirds of Earth's land surface. An insoluble problem over thousands of years, desertification has led to the abandonment of many cities, mass emigration, and increasingly frequent and severe floods and droughts. It is also a major contributor to climate change and a global threat to civilization.

[2] Royal game were uncommon and/or endangered animals such as sable antelope. They could not be hunted for any reason.

The thinking here was that if the landowners owned the game, they just might look after it. This I felt was sound thinking and fortuitously coincided with my game ranching objectives.

As mentioned in my Northern Rhodesian Game Department days, I had conceived of the idea of wild game harvesting or "ranching" when I first met British ecologist Frank Fraser Darling (or FD). My idea at the time was to put an end to the rampant commercial poaching of the aquatic antelope—the black lechwe—in the vast Bangweulu Swamps. Poachers drastically reduce animals when they use snares, dogs, and spears because they kill mostly females and young. However, controlled hunting, in which people harvest only adults and mainly males, increases the population.[3] I believed we could replace illegal poaching with legalized and well-managed hunting. This would thus save the lechwe and simultaneously provide food and income to local communities living alongside the lechwe. The principles on which I was working would allow the animals to exist *in situ*—in their natural habitat—there would be no fences and no feeding; they would remain wild, and their habitat undamaged.

Sixty years later, in writing this book, it's clear my thinking at the time was hopelessly naïve. The "science" of selecting and harvesting wild ungulates (hoofed mammals) for efficient and sustainable commercial production does not differ radically from principles that applied to their domestic cousins, cattle and goats. And the idea that some ineffable quality of "wildness" conferred on game a mysterious capacity to thrive without habitat damage, whereas cattle and goats predictably destroyed the habitat, was in fact based on ancient and unchallenged beliefs.

But I was young, aspirational, and figuring it out—including being a father. I was due to take some leave, having never really taken a break over the past five years, and Shirley deserved one, too. We planned a leisurely drive down to the sea in Natal, South Africa, where Shirley's family lived. They would all meet newborn Megan, Jim would have the opportunity to dip his toddler toes in the sea, and Shirley and I could get much-needed rest and relaxation.

Alas, it didn't happen.

The year before (1959), while with the tsetse department in the Zambezi Valley, I had heard of two American wildlife biologists who had come to Southern Rhodesia under the auspices of the Bulawayo Museum. They were Fulbright Scholars Archie Mossman and Ray Dasmann (soon nicknamed Mossie and Dassie[4]) working at the museum with Reay Smithers, an old family friend who had allowed me to practice taxidermy at the museum as a schoolboy. Archie and Ray had come with an idea like the one I had pitched

[3] It may be helpful to think of the concept of controlled hunting via the lens of cattle ranching. Despite slaughtering millions of cattle, ranchers' herds do not go extinct because the rancher is controlling what is killed, which is primarily the males and old or unproductive females.

[4] A mossie is the local nickname for a scaly feathered finch. A dassie (pronounced *dossie),* common in Southern Africa and known as a rock hyrax, resembles the American prairie dog but is approximately the size of a rabbit.

to FD and was now making a priority in my role as game research officer: If game had a commercial value, maybe game ranching, in which we harvested wildlife and fed people, would create an incentive to save it. Game ranching could in turn replace cattle, which were considered damaging to the environment.

Archie and Ray proposed working with the game department to see if they could get game ranching going in Southern Rhodesia. From everyone I talked to in the department, however, I heard only ridicule for these "interfering Yanks who knew nothing about African game." At one point, while still in the tsetse department, I learned that Archie and Ray were visiting a game department camp at the new Kariba Dam on the Zambezi River.[5] I immediately took this opportunity to meet the two scholars. Both men had studied in the United States at the University of California, Berkeley under the renowned conservationist Aldo Leopold's son, Starker, and they had connected with FD through his interest in managing deer populations. Upon meeting Archie and Ray, I dispelled the petty remarks I had heard about them and became determined to work with them should I get a chance. Later, when Shirley asked what I thought of them, I told her they knew more about game management than all of us in Northern and Southern Rhodesia put together—and I was going to find a way to work with them and learn.

I knew Archie and Ray were working on a game ranching project for the Southern Rhodesian Game Department under Graham Child and were currently stationed at Doddieburn Ranch.[6] Shirley and I were going to pass right by the ranch, enroute to our family vacation, so we decided to stop in briefly. I wanted to reintroduce myself and let the two men know I was now in charge of game research for the country, having taken over from Graham, and was looking forward to working with them.

Upon pulling into the ranch, I was shocked to find two dejected men holed up in accommodations graciously provided by the ranch owner, Ian Henderson, but without much else. Archie and Ray explained that for the past year, the Southern Rhodesian Game Department had delivered nothing but excuses for the lack of support. The scholars had requested tents, rifles, and other supplies, only to be told that none were available. To the contrary, I knew our stores were full of supplies; I was ashamed of how my department was treating such important scientists. I simply had to stay and get their project started. Shirley kept a stiff upper lip and went on to South Africa with Jim and Megan to visit her family.

It took one irate phone call to headquarters to get Archie and Ray all the equipment they needed. I had my Land Rover driven down to me from Salisbury and stayed with them for the next three months. We got game ranching started, first on Doddieburn and

[5] At the time, this camp is where Game Ranger Rupert Fothergill was mounting the rescue operation, Operation Noah, which I describe in more detail in Chapter 15.

[6] Doddieburn Ranch was situated south of Gwanda, a small town between Bulawayo and the South African border.

then on other ranches in the southern and south-eastern lowveld. To help Archie and Ray with local knowledge, I recruited my friend Peter Johnstone, who had hunted with me in my childhood and university years and was now with the tsetse department. Ian agreed to employ Peter as his first game ranch manager. Peter would receive training on the job.

In the early 1960s large areas of Southern Rhodesia were designated state land. This land consisted primarily of unsettled tsetse-infested areas, game reserves, and state forests run by a quasi-government forestry commission (for which I had worked as a student). The remainder of the land fell into two sectors: deeded commercial farms and ranches, mostly owned by White people; and native reserves (communal lands) administered by the Native Department under which were the traditional chiefs and headmen. Since the native reserves—which had more donkeys, goats, and cattle than people—were largely devoid of game other than crop-raiding elephants, we first focused on establishing game ranching on deeded cattle ranches. This is not to say that all classes of land were not losing biodiversity and deteriorating.

One of the next ranches to get up and going with game ranching was Buffalo Range, owned by George Style. The ranch lay southeast of Doddieburn, near the town of Chiredzi. George was a family friend—I had gone to school with his sons, Rodney and Clive, who helped him run the ranch and a store. On George's ranch we did our game surveys using the strip-counting technique,[7] which usually underestimated wildlife populations but was a sound place to start. More importantly, we did sex and age classification sampling[8] to get an idea of population health for each species. Based on this information, I submitted a report to the Style family, outlining the quota of game they might harvest. I also sent a copy of the report to the game department so they could issue their permit.

About six months later I was passing by George's ranch and stopped in to see how things were going. I found a newly built butchery equipped with cold room facilities, a skinning shed, and all that was required to operate, but no activity. The family informed me that they were still waiting for their permit! I told them to proceed using the quota I had determined and rushed back to Salisbury, furious.

Back at headquarters I found my report under a pile of papers in a clerk's in-tray. The clerk tried to tell me that, over the past six months, he simply had not had time to issue the permit. Incensed, I shared a few terse words with the clerk, emphasizing that we were civil servants, there to serve the public. Honest citizens had invested considerably on our

---

[7] In the American-developed strip-counting technique, two pieces of information are recorded: the number of animals seen on an area of sampled land and the estimated vertical distance from the animals to the road or track. From this data, total numbers are roughly estimated.

[8] To do sex and age classification sampling, we tried to determine how many females had given birth (whether to twins or not) and what percentage of young males and females from the previous season's breeding had survived to breeding age.

advice and were now sitting around, not earning income, because some officious clerk was too busy drinking tea (they drank a lot of tea in the game department)! It took him ten minutes to issue the permit.

Developing game ranching under the aegis of the Southern Rhodesian Game Department involved a paradox that I did not fully understand until much later in life. Institutions, once formed, take on a life of their own, often going against their very purpose and mission in an effort to grow, finance, and protect themselves. I was soon to learn that my game department, the agency whose main purpose was the protection and preservation of wildlife, had little enthusiasm for protecting and preserving wildlife, nor serving the public, no matter the law. Rather than trust and encourage the public, the department expressed mistrust and exercised control through policies like issuing permits. Though nominally "giving ownership" of the game to the landowner, the department still "owned" the power to determine what animals could be harvested, when, and how. The department would also profit by charging stiff fees for permits, revenue that never went to wildlife management but instead back to the treasury to reinforce the idea that we were a revenue-earning department. None of this was conducive to research or the training of staff to be more proficient wildlife managers, and inevitably conflict began to develop between Fraser and me. As director and a former accountant, Fraser measured his success by how much income he brought into the national treasury and by the growth in numbers of his staff.

I convinced Archie and Ray, as well as Ian, George, and other ranchers who had started game ranching, that they should form a game ranchers' association. Accordingly, I orchestrated a meeting to be held at one of the ranches and had the ranchers invite Fraser and me, only after the agenda was set. At that meeting, Fraser could not object to a *fait accompli,* and the Game Ranchers' Association was born to represent and protect their interests.

Amidst my game ranching initiatives (which I hoped were paving a path to reversing the destruction of habitat but were still in early stages and unproven), I was becoming increasingly alarmed by the environmental degradation occurring in the wildlife areas that had no livestock, such as the lower Zambezi Valley in the north. In every case, officials blamed environmental degradation on there being too many wild animals and/or too little rainfall to sustain ample growth of forage. But in most of the country, where livestock rather than wildlife was run, the land was also rapidly deteriorating. Southern Rhodesia's first national park, Wankie, wasn't much better. It was indeed a confusing situation. In my mind, there were more questions than answers; we had hardly begun to identify the problem.

I could not help but notice one strange contradiction amongst my scientific colleagues. In native reserves, they claimed the deterioration was due to too many cattle and goats, but in commercial ranching areas they claimed it was due to the growth of invasive brush and trees. One question I often asked officials was, "Why are you fellows promoting brush clearing in the commercial ranching areas?"

The stock answer was, "Because there is so much invasive brush."

To which I would ask, "What is causing this?"

The stock answer was, "Too much overgrazing."

I would then ask, "Why aren't you advocating brush clearing in the native reserves?"

The stock answer was, "Because it isn't a problem there."

To which I would ask, "And why is that?"

Their reply, "Because there is so much overgrazing."

And to this I would just shake my head, saying, "You fellows need to be thinking more deeply because giving the same reason—overgrazing—for both an excess of invasive brush on ranches and an absence of invasive brush in native reserves defies logic."

The fact that this muddled logic was being advised by officials deemed experts, who were boasting esteemed scientific training, did not help matters. I often wondered why they didn't seem to be seeing what I was, and how their views could differ so much from mine (even though I was not fully understanding the degradation). I attributed it to me being an ecologist with absolutely no training in veld (or range) management, which was their training, but also to my training of army trackers, which sharpened my powers of observation (I dive into my army days in later chapters).

In 1961, I floated a proposal to Fraser and National Parks Director Les Stewart to finance a small team to take a year's sabbatical. The team and I would use this time to travel to other countries and see what solutions were being implemented, if any. Both Fraser and Stewart outright rejected my appeal.

I was yet to understand the assiduous lengths professional "experts" and particularly their institutions will go to avoid evidence that might expose their ignorance. Yet, I was soon to witness this, when I collaborated with both the game and veterinary departments on a report addressing the cause of the spread of foot-and-mouth disease. I had been tasked with drafting the report, which I did, indicating in two or three places that we just didn't have answers. When I shared the draft with my coauthor in the veterinary department, Dr. Harold Roth, whom I truly liked, he stormed into my office demanding we remove "all admissions of ignorance." When I asked what he would recommend instead, he didn't have a clue, emphasizing instead that, as a matter of *principle,* he would never put his signature on a report acknowledging ignorance. That was, as I recall, the end of my work on the report. Equally strong willed, I declared I would not sign a report in which we pretended to have knowledge we simply did not have.

By 1962, land in the southern lowveld well south of Bulawayo that once grew enough vegetation to support many thousands of animals could now sustain very few. When I went out to do game counts on the southern end of the enormous Liebig's Ranch, land once lush and thriving, I could barely find enough grass to feed my two horses. At the end of each day, my game scouts and I, on our hands and knees, foraged grass along the paved Bulawayo-to-Beitbridge highway to put in nose bags for my horses. There was no other grass for miles.

Close to Liebig's Ranch lay an odd semicircle of land jutting into Botswana from Fort Tuli, which was just inside Southern Rhodesia. This land, ten miles in radius, was known as the Tuli Circle (it was just beyond the range of canon fire, ensuring that shots fired from the fort, situated at the center of the circle, wouldn't fall on foreign territory). The wildlife in the Tuli Circle was managed by the game department and was in terrible condition. All the cattle in the immediate area had died, and a conservatively estimated 50,000 wild animals—from hippo and zebra to impala, kudu, and wildebeest—were starving and dying too. Upon witnessing this, I frantically tried to get permission from the department to at least shoot the animals in the worst condition, putting them out of their misery, but no permission would be granted. No matter what data I provided, Fraser always responded that he could not get approval for us to act. It was common to hear, "We're just not getting the rain we used to get; what else can be expected in a drought?" This belief in those days seemed to justify inaction.

In desperation, I turned to Archie (Ray had returned to his university position); as a respected American wildlife management professional, might he be willing to accompany me to discuss the matter in person with Fraser? Archie jumped on the opportunity.

Archie and I drove together to Fraser's office. We decided Archie would do the talking, so I would not appear insubordinate to my superior. Fraser invited us to sit down in front of his desk, but we chose to remain standing. Fraser opened with a hearty, "Well boys, what can I do for you?" Archie then calmly unloaded a litany of complaints about the terrible situation and the fact that the reports he had helped me write were getting no response. I listened without expression, even as Fraser feigned commiseration with us, emphasizing that he *wished* he could do something.

Then Archie asked, "Well, what have you tried to do?" Fraser said he'd tried everything he could think of, but the "powers-that-be" would not approve. Archie sprang like a leopard. "Who are these 'powers-that-be'? May we see the reports that were sent and the responses?" Fraser kept trying to back Archie down, but Archie persisted.

Finally, Fraser leaned back in his chair, threw up his arms in surrender, and declared, "All right boys, you've got me! I have not done anything because it is just too political." We shook our heads and walked out. Once more I was seeing that the game department's "priorities" had little to do with protecting and preserving wildlife.

CHAPTER 15

TOO MANY OR TOO FEW?

Counting Raindrops . . . and Animals

1962–1963

As humans, we see what we believe sooner than we believe what we see.

A sable antelope, whose numbers were steadily decreasing even though they were protected by the government as "royal game" (not to be hunted for any reason).

Some good was to come from the dreadful loss of wildlife in the Tuli Circle, though it was accompanied by a serious ruffling of feathers. Not convinced that the rains had dramatically changed, I launched into action. I reviewed many years of rainfall records from the recording stations, which showed low and highly erratic rainfall with runs of successively wetter or drier years. I could not identify the terrible droughts that were being blamed for the environmental degradation and concluded that the problem was not the rainfall, but the soil's inability to absorb the rain. The recorded precipitation ran off or evaporated before the plants could use it to grow or deeper soil layers could store it. No one was taking into consideration the bare, crusted-over soil between the grass plants, a degradation so extensive that in some areas more than 90 percent of the soil appeared to not be absorbing rainfall, although superficially the grassland appeared healthy.

The Tuli Circle was situated in the headwaters of the Limpopo River System, which arose in Botswana, South Africa, and Southern Rhodesia. In 1961, the governments of all three countries were providing drought relief aid to the farmers in the upper reaches of the Limpopo. At the same time, the International Red Cross was collecting money on the streets of Johannesburg and Bulawayo for flood victims in the lower reaches of the Limpopo in Mozambique. To me, this juxtaposition only made scientific sense if normal, albeit erratic, precipitation in the headwaters was no longer soaking into the soil but rushing off as flood water. Technically speaking, that would not be a drought; it would instead be "noneffective rainfall" and a result of human mismanagement. The drought was not causing the land degradation nor the deaths of thousands of animals, but our management of the land and animals in the Limpopo's catchment area was causing both the "drought" and the "flood."

The term *effective rainfall* had long been used by engineers in relation to crop production, especially under irrigation. It referred to the amount of rain that was usable by the plants. But this definition was meaningless on the many millions of acres that made up the watersheds and river catchment areas. I redefined *effective rainfall* to better express what was happening in these areas. Effective rainfall is rain that *soaks* into the soil and only leaves it by transpiring through plants or by flowing *through* the soil into rivers, wetlands, and underground reservoirs. This can only occur when the soil surface is covered by plants or plant litter. Most rainfall, I found, was "noneffective" in that it ran *off* the soil surface—resulting in floods—or evaporated back *out* from the surface, resulting in "droughts."

When I began to speak publicly about drought not being the cause of land degradation, but land degradation being the cause of most droughts and floods, I not only agitated my fellow scientists in government, but also scientists in South Africa. They were so convinced that increasing droughts were responsible for land degradation that they arranged an international conference in Bulawayo, chaired by a Dr. Pereira from South Africa, to present papers analyzing the rainfall data as far back as the earliest records. But none could establish any run of droughts. They could only confirm the highly erratic nature of the rainfall.

While I was confident that land degradation was due to our management, not rainfall change, I fell short when it came to recommending what to do about it. If it was not drought causing the problem, then it had to be due to too many animals. And if that was the case, then presumably we had to figure out an ideal number of animals on the land and take action.

A major project going on in the Southern Rhodesian Game Department at the time gave me an opportunity to better understand the value of game counts. Coined Operation Noah, the project entailed rescuing wild animals from the islands created as Lake Kariba filled behind the newly constructed Kariba Dam on the Zambezi River. Though Operation Noah was essentially a publicity stunt (most of the animals could swim from the islands to the mainland and did so on the northern side where the Northern Rhodesian Game Department did not mount a similar operation), the project succeeded in turning attention away from the terrible international image we had earned through our tsetse fly eradication program. The project, under Game Ranger Rupert Fothergill, was credited with rescuing more than 6,000 animals, from lions to elephants to warthogs.

Though I had no official role in Operation Noah, the counting of wild animals was an important part of it. Hence, I was a periodic visitor, gathering data and often volunteering my assistance. For each island in Operation Noah, Rupert and his experienced team had to first estimate the wild animal numbers, then the time, labor, fuel, food, and equipment needed to rescue that number.

On one thirty-five-acre island, I participated in the wildlife count as one of Rupert's team of forty men—more than one man per acre. Counting the game was easy on this island because most of the leaves, which typically concealed the animals, had been eaten off by the buffalo, kudu, and impala. There were three baobab trees which men climbed and surveyed from, like silent stationary helicopters. The rhino count was accurate as was that of the few buffalo. The counts of most of the animals, however, were astoundingly inaccurate. Sixty kudu to be rescued turned out to be 120, and 150 impala to be rescued turned out to be more than 300.

My mentor, British ecologist Frank Fraser Darling (or FD), had personal experience with the problem of estimates as well. During one of our fireside chats in Northern Rhodesia, he recalled his experience counting deer in a forest in Scotland. The local people and game keepers "knew" there were twelve deer, pointing out that the deer were quite tame and frequently seen. FD decided to check that count by having a large group of men and dogs move through the forest and drive all the deer out. He posted observers to record every deer emerging as well as its position and the time to avoid the same animal being counted twice. The twelve deer turned out to be forty-eight. To this day I note that people consistently spend enormous amounts of money and time estimating animal numbers, believing that the total number is critical to management plans.[1] But I was soon to realize that I no longer shared this belief.

[1] At the time (and still to this day) both game harvesting and safari hunting permits were based on estimates of wild animal numbers. I didn't trust the accuracy of even the most respected methodologies (those developed in the United States from the trailblazing work of the father of game management, Aldo Leopold), but they were the best we had.

As the nascent game ranching industry grew, I got Director Fraser's permission to train a few game rangers to assist me in supporting the ranchers. Though I included the how-to of estimating wild animal numbers by making sample counts on the ground or from the air, I also covered how to do sample counts based on wild animal signs such as tracks and droppings along fixed paths or transects. And now, after having worked with scholars Archie and Ray for many months, I shared ways to sample the sex and age ratios (females to males, young to old) in the various animal populations. From this data we could derive the proportion of females surviving to breeding age each year. If the frequency of tracks was increasing and there were more young females surviving to breeding age, we could be reasonably sure that that animal's population was increasing, although we had no idea how many animals there were. I also trained the game rangers to recognize when grass plants had been overgrazed or brush and trees overbrowsed,[2] and whether young browsing animals were likely to be malnourished because adult animals had browsed the feed within their reach. At the end of the training, I asked the trainees to list every indicator they could think of that might help interpret the health and trend of various wild animal populations. Their lists were as long as twenty indicators, including of course animal numbers.

Then, I tasked the trainees with ranking the indicators. If they could only ask one question, what would it be? What would they want to know next? And so on, down the list. Without fail, when all the questions were prioritized, knowing the actual animal count landed near the bottom of their lists. We were spending the most money and time gathering information, upon deeper analysis, on what was perhaps the *least* important information needed for good management! Clearly, we had a major problem accurately counting wild animals in thick bush, but that still didn't mean that there weren't too many if we observed damage, or did it? Around and round my mind spun.

While Operation Noah was going on, I became increasingly concerned with the area downstream of the dam in the lower Zambezi Valley (now Mana Pools National Park); it was suffering some of the worst degradation within my jurisdiction. However, as I had experienced, convincing people of a problem when they are oblivious to it is no small task. I knew I would first have to gather data and document the damage before even having a chance to discuss management plans with Director Fraser. I was, after all, in the business of research.

I forged ahead on the research for yet another report. What I did not recognize then was that I was now back to where I had been during the long weeks of campfire discussions in Northern Rhodesia with FD. We held the same deeply engrained beliefs as our colleagues about what we were seeing. If vegetation was disappearing when grazed or browsed by animals, then there were too many animals. Not only was that the accepted

[2] Grazing refers to feeding on grass plants, while browsing refers to feeding on woody plants—shrubs and trees.

wisdom of thousands of years but it was the teaching of our universities and appeared to be "common sense." So now, the only factors I could see that were damaging the land was our burning of grass, which exposed soil, and an overpopulation of animals eating everything in sight. As my university education had confirmed, if overbrowsing or overgrazing were occurring, there were too many animals, and I confidently (albeit naïvely) set out to show with solid data that this was the case in the lower Zambezi Valley.

The research division I took over as game research officer included an empty office and meager budget to cover fuel as well as my game scout, Scarface's salary. I needed photographic records and to do so, had to buy a camera and film with my own money. I also needed a darkroom to develop my film and make prints. After Shirley ruled that on top of handling two children and hand-washing the laundry in a half-built home, she was not turning our bathroom into a darkroom, I found a vacated storeroom at the office to do the job. There, I also housed a library and a punch card filing system for the data we gathered.

Tasked with an immense amount of ground to cover, no research staff, and no resources, I relied a lot on what I could interpret from limited information. Rupert Fothergill had reported finding a set of reedbuck horns at Mana Pools, and I had recorded the last female seen alive downstream, at the mouth of the Chewore River. Sable antelope were rarely seen and clearly falling in numbers with few reaching breeding age (in 1930 my father had reported sable as some of the most commonly sighted animals while he was surveying a potential rail line through the Zambezi Valley). I documented the disappearance of plant species such as *Ficus caparifolia* (a rambling fig), and *Oncoba spinosa* (snuff-box tree), as well as virtually all reedbeds along the south bank of the Zambezi River. I noted the lack of seedlings and young trees in major riverine tree populations.

In my report I concluded that the debarking of mature trees was due mainly to elephants; the lack of seedlings and new trees due to overbrowsing by buffalo, impala, eland,[3] and kudu. I recommended that the elephants and buffalo, which were in the highest numbers, be reduced to a number the land could more likely sustain. This meant culling these populations before the Zambezi Valley mirrored the Tuli Circle and its devastating loss of wildlife.

I finalized my report and sent it to Director Fraser. As expected, the very idea of government shooting animals in any game reserve or future national park immediately raised alarm; I once again heard him cite my recommendation as "so political." Accordingly, Fraser ordered me to rewrite my report, which I duly did, changing the English but not my recommendations. He then ordered me to write the report again. I repeated the process. Each time, Fraser would not say why I was being told to rewrite it, but I knew

[3] An eland is the world's largest antelope and looks like a muscular cow; both males and females have thick, twisted horns and run in sizeable herds.

damned well he wanted me to simply report the data and leave the management recommendations up to him. The third time he asked for a rewrite, I refused. It was then that Fraser ordered me to omit my recommendations, sternly reminding me, "Savory, you are a research ecologist and should only be providing data."

I replied back, just as sternly albeit not tactfully, "I am employed as a scientist, not a prostitute!"

The situation was resolved only when government formed a committee under Reay Smithers, director of the Bulawayo Museum. Smithers recruited several respected Rhodesian scientists, including Dr. Oliver West (our old family friend who was now chief of Botanical and Ecological Research in the Ministry of Agriculture) and Dr. Fido Phelps (the entomologist I had previously worked under in the tsetse department). The committee visited the field with me, where we held many constructive discussions. On viewing the evidence, they supported my recommendation: *Animal numbers needed to be controlled to prevent the environmental damage occurring in our game reserves and future national parks.*

Despite the outcome of the committee's investigation, the culling of excess numbers of animals did not happen until after I had left the department. Eventually the game department and national parks department came under professional management and were combined into one (National Parks and Wildlife Management Department), headed by my predecessor Graham Child, who by then had earned his doctorate.

In subsequent years Child embraced my recommendations, and national parks staff shot more than 40,000 elephants in culling operations. My misinterpretation of the data to fit the beliefs of the day led to the slaughter, which has haunted me ever since. As humans, we had all chosen to see what we believed sooner than we believed what we could see. We had interpreted the data about destruction of vegetation and soil cover to fit the centuries-old belief that *if animals are causing damage, it is due to too many of them.* I would eventually completely refute this belief, but in the meantime I needed a few more eye-opening shocks to my system before I could truly see evidence of the real cause of the degradation, which was neither a change in rainfall nor too many animals.

CHAPTER 16

IN SIGHT . . . AND OUT OF SIGHT

Encounters with Painted Elephants and a Skilled Poacher

1962–1963

I circled the pool, searching, but the man had simply disappeared. Logically, he had to be in the muddy water but where?

Some of the "Zambezi Ladies"; trying to mark them with paint bombs was a miserable failure.

Of my repertoire of experiences while working for the game department, there were many that required me to carefully observe, to ensure I was not just seeing what I wanted to see, to probe deeper, and to question, question, question. There were also instances in which sight played a significant role in my findings in ways that seemed to have a surprising twist to them.

One of my most memorable experiences—which upon reflection was certainly the universe cautioning me to not "see what I believe"—happened right before I transitioned from the tsetse department to the Southern Rhodesian Game Department. The handful of us living on the Sabi River collected our mail at the nearby Chisumbanje Agricultural Research Station run by a fellow named Bob Mowbray. During one of my visits, I found

Bob down by the river, his dogs incessantly barking at a hole in the bank. The hole appeared to be that of a leguaan (monitor lizard), however the angry staccato sound coming out the hole sounded much more like a mongoose[1] when cornered. I suggested Bob pull it out, but fearing being bitten, he passed the task onto me. I knew that to minimize the chances of getting mauled, I would need to grab the mongoose by the throat, from in front. To do so, I first needed to get a good look at it. I peered into the hole, waiting for my eyes to adjust to the darkness. The rapid-fire threatening sound increased, and my eyes gradually made out two beady eyes staring back at me, about eighteen inches away. The eyes seemed too close to each other to be those of a mongoose. I asked Bob to look—what did he think this creature was? Albeit a bit more hesitant, Bob repeated what I had done, peering into the hole until his eyes adjusted to the darkness. At this point the animal's threatening grew furious. Bob agreed that its eyes did seem a bit too close to one another for a mongoose. As a caution, I asked if he had a torch, which he left to retrieve, because I wanted to confirm what it was I was about to drag out of its hole.

When we shone the torch into the hole, we were horrified to find ourselves face-to-face with a massive cobra! Until then I had no idea snakes could be so vocal and make a sound like an angry mongoose.

Years later, when horseback riding with a farmer one morning in Wedza, a few hundred miles north of the research station, his dogs started barking madly. When we cantered up, I heard the same "threatened mongoose" sound and shouted at the farmer to call his dogs off, quickly! Sure enough, the dogs had cornered a large cobra.

Back down in the Zambezi Valley I learned a good lesson in "seeing" from a poacher. I was driving in an open stripped-down Land Rover with two game scouts when we spotted a poacher in the open floodplain. The poacher, carrying a muzzle-loading gun and a small canvas pack, broke into a run and we jumped out to chase him. The two scouts could easily outrun me, and it didn't take long for all three to get far ahead. At one point I saw the poacher dip down into a hollow, followed a short while later by the scouts. However, only the scouts reappeared on the other side, no longer in pursuit of the poacher but running on blindly—they had, strangely, lost him. I stopped and picked up the poacher's tracks, which trailed into a pool of water in the hollow with no significant vegetation or cover. I circled the pool, searching, but the man had simply disappeared. Logically, he had to be in the muddy water but where? Baffled, I stood wondering how he had simply disappeared. It was then that I heard a faint sucking sound. Moving ever so slowly, I inched toward the sound, stopped, and listened. When the sucking sound happened again, I inched closer. I repeated this until I was eventually standing almost at the source of the sound, but still no poacher. Then, from the muddy water, the poacher's

[1] There are many species of mongooses; they are typically short-legged animals with pointed noses, small ears, and long furry tails. Although carnivorous, they typically try to avoid encounters with humans.

lips emerged with barely a ripple, exhaling and sucking in air. I stepped on his lips, forcing the poacher to fully emerge, now dripping in mud.

I marveled at this poacher's skill. He had managed to enter the water without a trace, submerge himself as well as his gun and pack, and remain face up yet out of sight, all with amazing speed while we were in hot pursuit. Eventually the forlorn scouts returned and, upon seeing the defiant poacher, began to verbally abuse him, angry that he had outwitted them. I stopped them, still in awe of the poacher's cunning. We brought the poacher back to our camp, where I chatted with him, admiring the extreme skill of his disappearance act. Priding myself on bushcraft, I challenged him to a fire-making competition using nothing but two sticks. He beat me.

The poacher's skills earned my respect and—angry at my scouts' abusive behavior when he outwitted them—I decided to let him go. I stripped him of his gun—which was punishment enough for a noncommercial village hunter—gave him a bucket and told him to go fetch water from the river. I asked the scouts to accompany him to ensure he did not escape. Just as I speculated would happen, two shamefaced scouts returned without the poacher.

Regarding how what we see—and don't see—can certainly change the circumstances, I acquired a few relatable skills on a zebra hunt where I accompanied Abe Dennecker, the top hunter in a biltong[2] business owned by two men, Glover and van Baart. Glover and van Baart had been operating their successful biltong business on Liebig's Ranch. Zebra and wildebeest were considered vermin in that region because they competed with cattle for grazing, and hence ranchers had allowed the men to hunt them throughout the year. My purpose in joining the hunt was that, in the interest of developing game ranching, we were seeking efficient and humane ways to harvest animals. I wanted to learn what these most experienced professional hunters were doing to be so successful in harvesting the wildest of animals.

I had heard that Glover and van Baart's hunters mainly ran down zebra and wildebeest rather than stalk them, the latter being standard practice for most hunters, including me.[3] Abe briefed me on their technique. When it came to hunting zebra and wildebeest, he explained, his hunters and he only stalked as close as they could for the first shot; from there on it was a running hunt. If a hunter didn't run directly behind fleeing animals but rather *at an angle,* the animals would unwittingly circle to keep him in sight. This was new to me.

Early the next morning off we went. We drove along various bush tracks until the tracker, sitting on the front of Abe's short-wheel-base Land Rover, picked up fresh tracks of

---

[2] Biltong, popular in Southern Africa, is made from the best large muscle cuts, brined, and shade-cured to create a taste only comparable to *Jamón Iberico* from Spain. The fact that it can keep for decades is part of its notoriety, from being the main sustenance for Afrikaners in the Boer War to the main protein provided to British polar explorers. Biltong is not to be confused with jerky, which is sun-dried thin cuts of meat.

[3] The only quarry I would run down was dangerous game that I'd wounded. If a lion, buffalo, or elephant didn't drop after my first shot, I would run behind the wounded animal, keeping it flustered, confused, and in sight.

a herd of eight zebra. We proceeded the rest of the way on foot. The trackers were excellent and spotted the skittish herd before being seen. Abe then stalked as close as he could, with me following. But before Abe could shoot, we were detected, and the herd took off at a gallop. Abe was unphased, quickly sprinting after them, motioning for me to do the same. As heeded, we were careful to run at an angle to the zebra, not directly behind. Sure enough, the herd curved our way, trying to keep us in sight through the scattered brush vegetation. Each time Abe had a zebra clearly enough in sight, he would shoot. The herd would again run away with us once more running at an angle to the side. Abe was fast and accurate. After he shot three zebra, I joined in, yelling as we ran, "I'll take the left most zebra, so we don't shoot the same one!" The technique was indeed efficient—in staying at an angle, the zebra kept turning our way. We downed all eight zebra within minutes. In seeking more efficient and humane ways of killing, an understanding of the inclination of zebra to keep us in sight proved invaluable for the hunter, albeit unfortunate for the hunted.

Learning on the job was standard, as almost everything I had to do as a researcher was in the early days of wildlife management before many of the techniques we take for granted today had been developed. To understand habitat deterioration in the lower Zambezi Valley I first studied vegetation damage, recording age classes of trees, the degree of damage, and what animals had caused the damage. When so much seemed to be associated with elephants, I needed more information on them, particularly their movements. But to successfully gather this information I also needed to easily identify which elephant I was seeing. How could I effectively mark individual elephants? This was indeed a dilemma, because the use of drugs and dart guns to tranquilize animals, not to mention radio collars and microchips, were far in the future and my research equipment budget was zero dollars.

I decided to try marking elephants with a long-lasting paint. I chose a durable paint used by the roads department that came in white and yellow. The department kindly gave me a supply but no advice on painting wild elephants, in this case the Zambezi Ladies[4]—by reputation the wildest of the wild elephants. I had had a close call with the Zambezi Ladies just months prior. My friend Clive Howard-Beard and I were observing a group of about fifteen of them near the Rekomitjie-Zambezi junction in open country scattered with large winterthorn trees.[5] Suddenly a group of cows with calves caught our scent and immediately charged us. I had my .470 but Clive had no rifle, so I told him to get behind a large tree, warning him not to run, whatever he did. My heart pounding, I then stepped

[4] The Zambezi Ladies were coined so by one of the most experienced ex-elephant hunters and poacher-turned-game-rangers, Ian Nyschens. Ian left the Southern Rhodesian Game Department as I joined it, so we never met. In his book, *In the Months of the Sun,* he describes what he called the Zambezi Ladies as the wildest, most dangerous elephants he'd ever encountered.

[5] Winterthorn trees, *Acacia albida,* are today known as *Faydherbia albida.* These large trees grow across Southern Rhodesia to North Africa, generally on flood plains. The common name, winterthorn, derives from the fact that when other trees lose their leaves in the dry, winter months these come out in full leaf. Each tree produces over a ton of high-protein pods that drop in the late dry season, making them desirable by a lot of wildlife, including elephants.

out into the open, hoping the matriarch would come straight at me, from which position she would be easiest to kill. And that she did. I shot her at the last possible minute, the rest of the herd thundering by both sides of us—Clive protected by the tree, and I by the dead cow's body. Had I shot the cow while too far from me or had the herd's concerted charge occurred in dense brush, I doubt we would have escaped.

The Zambezi Ladies spent much of their days lounging in very dense jesse bush,[6] hence I decided to try using a bow and arrow to hit them with my newly fabricated "paint bombs." My paint bombs were various sizes and shapes of light bulbs that I filled with paint. A friend in the workshop of the prisons' service created fittings that allowed me to attach the bulbs to the arrows. The idea was that when shot, the bulbs would smash against their target, leaving a good blob of paint.

In the field, all my paint "ammunition" failed. The bulbs would neither break nor the paint splatter upon impact with an elephant's flexible hide. Instead, the arrows would simply fall to the ground, only succeeding in annoying the elephant. In some cases, the bulbs hit branches and shattered, painting the vegetation yellow or white. The trajectory of my arrows was also pathetic, so I had to get extremely close, which certainly wasn't comfortable considering the whole scene quickly agitated the Zambezi Ladies. Eventually I discovered that if I just threw the bulbs by hand, with a good spin, they would smash on impact with an elephant's skin.

On one occasion, walking back to camp, I encountered a single bull elephant lying asleep on his side, his head against a winterthorn tree trunk. This was going to be a piece of cake! My game scout Scarface handed over a couple of bulbs from his pack. I put a spare in my rugby shorts pocket, and with my .470 rifle in my left hand and a paint bulb in my right, quietly snuck up to within two paces of the bull's backside. I threw the bulb with all my might at the bull's hip—and missed. The paint bulb went flying just over the slumbering bull and smashed into the tree his head was against. The bull jolted awake, lumbering to its feet. I put my hand in my pocket to deliver the spare. Just like the old "baboon and calabash trap,"[7] I could not get my hand out of my pocket while clutching the bulb. I had to push my hand deeper and let go, all this causing me to do a bit of a dance with time certainly not on my side. Fortunately, elephants tend to turn on their back feet, so I stayed out of sight behind the bull until he rushed away, leaving me with a thumping heart and the smell of road paint.

Eventually I simply abandoned that whole project as both impractical and dangerous.

My efforts to better "see" amidst my duties sometimes completely eluded me, and in

---

[6] Jesse bush refers to large areas of about twenty-foot-high dense bush, predominately *Combretum celestroides,* impenetrable except along elephant, rhino, and buffalo paths.

[7] A "baboon and calabash trap" is a way of catching baboons by placing food inside a hollowed-out gourd (calabash) with a narrow entrance. When the baboon reaches its hand in to grasp the food, the narrow opening prevents it from easily getting its hand out without relinquishing the food, effectively trapping it.

one such case, gave me more insight into the powers of racism at play. This time I was on the border with Mozambique, where the railway line crossed near Malvernia. I had stopped to fuel my Land Rover when a fellow noticed its game department emblem and approached me. The man introduced himself as a railway ganger and asked if I might be able to help with a problem elephant. The elephant, he said, was routinely raiding the gardens of his railway workers, terrorizing them all. Three times now one of our game rangers had shot "the elephant," but it was always the wrong one because the trouble continued.

I drove back with him, his workers assuring me that the bull would appear at about sunset. I figured that letting the bull charge me might be the best way to get the right elephant, and everyone agreed. So that I could see—at least vaguely—the end of my .470 barrel in the dark, I wrapped a strip of foil from a soup packet around it and started walking along the line. As darkness fell, I heard a rushing sound through the vegetation, and a charging elephant loomed before me. It was a clean shot, and the next morning the ganger's house servant graciously came to our camp bearing a gift from the ganger's wife—a tray with tea, milk, and cakes beautifully presented on crocheted doilies. I thanked her and enjoyed the confections.

After arranging for the ganger to send the tusks to the game department and take all the meat for his staff, my scouts and I began packing to leave. I soon noticed that the two scouts were uncharacteristically sullen. Clearly something was wrong. When I asked them about it, they denied there was a problem. The sullenness continued as we drove back to my camp. Finally, I stopped, got out, and demanded that they either explain their behavior, or I would promptly leave them there, still miles from home.

The anger poured out of them. Why had I accepted that tray of tea and cakes from a Colored ganger? Did he think he was White? Why did I treat him as though he were White by accepting that gift? I was completely stunned and calmly explained that I had always accepted and traded gifts with chiefs, headmen, and farmers, regardless of whether they were Black, White, or Colored (mixed race). My gesture had nothing to do with bloody race or tribe.

In later years I was to learn much more about the unjust middle ground the "Colored" population found themselves in, but for the moment it was a shock to realize how a simple gesture like accepting a gift from a Colored man's wife could mean so much more than it initially appeared.

CHAPTER 17

MY BROKEN BONES—AND FLAGGING SPIRIT

When Stress Becomes Evident

1962–1963

I came away with broken ribs and severely skinned legs that festered badly, but I was alive.

The eroding banks of the Zambezi at Mana Pools (1963). The soil-stabilizing fig thickets and reeds are gone, and so are the reedbuck and many other small mammals and birds. This tree would be lost, too, within a short time.

While on the islands with Rupert Fothergill's team during Operation Noah, I had my closest call ever with a young buffalo bull—and unlike other close calls with the wild, I didn't walk away from this one unscathed.

I had joined a crew capturing buffalo calves that Rupert thought would not be able to successfully swim with their mothers. The plan was to give chase to the herd on foot, catch the calves by hand as they lagged far behind their mothers, and tie their legs with rawhide rope. We would then release them on the nearby mainland. How the calves would rejoin their mothers was not a subject of discussion.

Our morning started off badly; as we were in chase, two buffalo bulls turned back on us, charging together. I had to shoot both, relieved that there were no other casualties—

just a terrified scout who had scrambled up a tree, seeking refuge. Yet the morning only got worse. As we resumed our chase—running through knee-high grass, trying the best we could to keep up with the herd—I saw the back of what appeared to be a half-grown buffalo running our way in confusion. The crew scattered but I put down my rifle, anticipating I could catch it. Oh, what a mistake!

What I thought was a small animal was a young bull, its size not apparent due to it being in a depression masked by tall grass. By the time I realized my mistake, the bull was almost on me. I desperately tried to grab his horns, but one hand slipped, and he hooked me under my left arm, tossing me over his back into the air. I hit the ground on my hands and knees, not able to get up before the bull descended upon me again. This time I managed to grab his horns and twist as hard as I could (something we had done as kids with oxen). He stumbled and came down, but my legs were now pinned under his shoulder. In a blind rage, he lifted his head and began bashing me on the ground. Each time he lifted his head I went up with it, hanging onto his head for dear life, my chest behind his right horn. If his horn striking the ground hadn't absorbed almost all the shock, I would have been mashed to pulp. Still, I could not get free and began to think I wouldn't make it out of this one alive.

Suddenly my game scout, Scarface—who I assumed had fled with the rest of the men—was beside me, yelling *Yegela! Yegela!* (Let go! Let go!). I did as he urged. As the bull rose without me, Scarface grabbed my arms and dragged me away to safety. Unbeknownst to me at the time, Scarface had used one of the rawhide ropes to ensnare the bull's back legs while it was focused on shaking me off its head. Scarface's quick thinking saved me; I came away with broken ribs and severely skinned legs that festered badly, but I was alive.

As a research officer, I needed to get into more remote areas—particularly in the lower Zambezi—so I stripped down my Land Rover and fitted it with an extra fuel tank plus holders on the front and back for four jerry cans of diesel (holding about five US gallons each). I also bolted metal army aircraft landing strip plates along the sides to deploy if the vehicle were to get bogged down in sand or mud. There was only one rough track down to the valley at the foot of the Zambezi escarpment which forked into two, one track going to the Sapi River and the other to Mana Pools. By using these two tracks and then slowly driving across the bush, my Land Rover could take me to within a day or so of walking to wherever I needed to go. With just Scarface as my assistant and no radio, we could not afford breakdowns and planned accordingly. When the temperatures would drop to near freezing at night, for instance, I took to draining the entire radiator and engine block in the evening, then filling them with hot water at dawn to ensure the battery had enough life to start a cold engine.

One day in the Zambezi Valley, at the start of the rainy season while I was ever so carefully driving up the bed of the Chewore River to avoid the impenetrable brush on either side, my Land Rover sank to its axels in the soft wet sand, stuck. Scarface and I got to work wedging the metal landing strip plates behind the back wheels. We needed a few short sticks to corduroy the sand behind the front wheels, which Scarface collected. I started cutting up the sticks with a panga (machete) when it unexpectedly slipped from my wet hand, slicing through my sandal, all but severing the big toe of my left foot. I slumped against the wheel in pain, bleeding profusely. Feeling faint, I realized I had to do something. I considered just cutting the severed toe off but while inspecting the toe, a piece of knuckle bone fell out. I decided to try to save it. I gingerly washed off the bone in the clear water flowing over the sand, put it back in place, and wrapped toilet paper around my toe to stanch the wound. Scarface then cut my shirt into strips, and we created a bandage as best we could.

With Scarface not knowing how to drive, the trip back took forever. I had to keep the Land Rover mostly in first or second gear because it hurt too much to depress the clutch with my left foot. We made it to Rekomitjie Tsetse Station, my old haunt, that night and stayed there, driving on at dawn. By the time I got to the hospital in Salisbury, my entire leg was throbbing. The doctors reprimanded me for putting the piece of bone back in, as it had in turn infected my foot. They operated on my toe, discarded the bone, and regrettably informed me that the wound would almost certainly heal rigid, causing a limp in my left leg for the rest of my life. Determined that such would not be my fate, and having already had my gait permanently altered by polio in my right leg, I practiced forcibly bending my toe as soon as the pain subsided, then worked on picking up a pencil with my toe, day in and day out. Today, thankfully, no one would know my toe had been severely injured.

In 1963, shortly after my last difference with Director Archie Fraser over the appalling state of the future Mana Pools national park, I received a phone call from my father. He wanted to let me know that I was to be fired on the recommendation of Director Fraser. In fact, I was to be fired immediately without any right to appeal. My father by then was the most senior civil servant in the Ministry of Water and Irrigation under which the game department fell,[1] and had been informed as a courtesy. I told him I had done nothing seriously wrong and advised him to stay out of the mess. I would deal with it myself.

Soon enough, Fraser contacted me but refused to state the grounds for my dismissal. He merely informed me I was being dismissed. I said I would appeal the decision to which he informed me that the Civil Service Commission[2] had agreed that he could terminate me immediately, without any right to appeal.

---

[1] The Southern Rhodesian Game Department was in the Ministry of Water and Irrigation because of the interconnectedness of wildlife and water resources in the country's management and conservation efforts.

[2] Southern Rhodesia, like most British colonies, had a politically independent Civil Service Commission (CSC) that included, among other roles, a source of appeal for victims of unfair practices.

Nevertheless, I wrote to the head of the commission, Buller McKenzie, requesting an inquiry into my dismissal. I argued that the matter was not as straightforward as it might appear, and I felt it wrong to assume that I was guilty by accusation alone. Buller granted me an inquiry, and Fraser was obliged to provide the written grounds for my dismissal. Fraser's account detailed a total of forty-two allegations against my character and conduct, but none against my work.

The court of inquiry was composed of three men from the Civil Service Commission, sitting as judges to assess the evidence via testimony from "witnesses" Fraser called to the stand. The chairman opened the inquiry by stating that matters were not as simple as outlined in the dismissal notice, and the roles of accuser and accused might change as the inquiry proceeded. I realized I was likely to get fair treatment.

I admitted that two of the forty-two allegations were true: first that I was critical of Fraser's leadership; second, that on one occasion in the office I had told his secretary—who later became his wife—to "behave herself" (which I had indeed done in response to a startlingly close encounter in my darkroom). We then suffered through two tedious days of deliberations during which Fraser repeatedly called witnesses who either outright dismissed his allegations or whose stories held no merit.

Toward the end of Fraser's presentation of the allegations, while the judges were on their tea break, Fraser approached me, asking if I was prepared to withdraw. I pointed out that I had nothing to withdraw from, having made no allegations or complaints. He was firing me, not the other way around. Fraser then asked, if he withdrew all complaints, and the court allowed him to do so, would I not pursue the matter further? I agreed to those terms.

When the judges returned, Fraser promptly apologized to the court, admitted his allegations against me were "false and written in a spirit of spite," and pleaded approval to withdraw all allegations. The chairman consulted his fellow judges, then asked me if I wished to pursue the matter further. I told him I had given my word to Fraser that I would not. He then ruled that all the records be destroyed and instructed me to be reinstated.

We resumed working together, this time under pleasantly changed circumstances. But the experience left me feeling like government service might indeed not be a fit for me. I mused on all the men I knew in both the Northern and Southern Rhodesian Game Departments. What did I truly have in common with them? A few of us shared a genuine love of wildlife and the African bush life, but most were motivated by their desire for public power, influence, status, praise, and the pleasures of "khaki fever" (as we coined young women's attraction to game rangers).

Fraser was soon after transferred "upstairs" in the ministry, and Needham Gibson, a very decent accountant with whom I got on well, replaced him as director. At the same time a new post of assistant director was created. I was bypassed for it, and Dr. Harold Roth (my old colleague from the veterinary department who had argued that professional people should safeguard their status by never admitting to ignorance) was appointed instead.

In addition to my disenchantment with the motives of my colleagues, I started experiencing health problems. I drove into the office day after day feeling fine, but within a short while I would have to rush off to the toilet and throw up. My doctor suspected work-related stress, and circumstantial evidence proved him right. My entire physical system was reacting to the pettiness of the larger system in which I resided. I was suffering from a clinical allergy to the behavior I experienced around me.

This was not how I wanted to feel. It certainly wasn't helping me focus on the most critical professional task at hand—solving the degradation that stemmed from our management of both wildlife and human habitat. What was causing the loss of biological diversity, of land turning to desert, and what could we do to reverse it?

To answer these questions, I realized I needed to free myself from the institutional and professional pressures to never admit error. I promptly resigned from the Southern Rhodesian Game Department, my illness ended abruptly, and I determined to never again work in an environment where I felt honest research was—at least for me—not possible.

PART IV KAPOMBE FARM

CHAPTER 18

AND THE WINNING TICKET IS . . . KAPOMBE FARM!

My Self-Employment Plan

1963

We named it Kapombe Farm, in memory of my loyal former elephant control guard Kapombe Mukumpulwa, the one who had been murdered by poachers.

The site that would eventually became our new home at Kapombe Farm, shown here with our equipment (1963).

The banana that leaves the bunch gets skinned, as we used to say, yet in resigning from the Southern Rhodesian Game Department, I vowed that no matter the tribulations, I would never give up on my quest to find solutions to the degradation occurring under what most believed was the best of management.

I felt renewed in now being an independent scientist, unrestricted by institutional behavior or academic disciplines. However, I immediately faced one giant hurdle. Until now I had focused solely on my calling, and I had never given a thought to salary or income. Now, with a family to support and ambitious plans to change the world, I hit the brick wall of reality: somehow, I had to pay for it.

Fully committing to game ranching with my friends was a possibility but it was all still experimental (of which I delve into more in Part V, Game Ranching). I chose to go into an older, and presumably proven, model of agricultural development: farming. To me, farming was a sensible plan to support my family and, ultimately, to save enough money so that I could apply to a doctorate program in wildlife management in the United States, which Archie Mossman was encouraging me to do. As part of this plan, I would provide consultation in game ranching on the side.

My decision to go into farming came about by a Southern Rhodesian government incentive to settle farmers on raw bushland that they would then develop into irrigated farms to grow sugarcane. Alongside Hippo Valley Estates,[1] these farms were to draw irrigation water from the newly constructed Kyle Dam. In addition to the water being guaranteed by the government, Hippo Valley Estates' sugar mill committed to buying the sugarcane from each farm. Young farmers were invited to apply; of the applicants, thirteen "winning" names would be drawn from a hat. A cousin of mine intended to apply, and my father said he was considering investing in my cousin's endeavor if he were successful in the lottery.

Until this point I had had no interest in farming; however, my desire to make an independent living that would support my scientific work—combined with what little confidence I had in my cousin's ability to manage a farm—brought my father around to agreeing to invest a major part of his pension funds in me if I "won" one of the farms.

There was one more hurdle for my father and me to surmount—who would be boss? It was especially important for me to determine who would have the final say should crisis befall us. My father said we would make major decisions together, but at the end of the day, he would have final say, having put up the money. I convinced him otherwise; while it might be his money, it would be my life and that of my children. My father looked hard at me, then extended his hand and we agreed with a shake that in any crisis I would have final say.

Luck was on my side, as my name was indeed drawn in the lottery, and I became the proud owner of an undeveloped bit of bush with forty-foot-high mopane trees[2] scattered with buffalo and kudu on the edge of Hippo Valley Estates. We named it Kapombe Farm, in memory of my loyal former elephant control guard Kapombe Mukumpulwa, the one who had been murdered by poachers. And so, my farming career began.

On paper these farms were prizes comparable to winning the national lottery. They ranged from about 350 to 450 acres with at least one-and-a-half cubic feet per second[3] of

---

[1] Hippo Valley Estates was a large sugar company in Southern Rhodesia's southeast lowveld with thousands of acres of sugarcane and a sugar mill nearby.

[2] Mopane trees *(Colophospermum mopane)* only occur in Africa and have distinctive butterfly-shaped leaves. Where they occur, they are often the dominant tree species.

[3] Cubic feet per second (CFS) is a unit of flow rate, specifically the volume of water that passes a fixed point in one second. It is a measurement of the quantity of water moving, not its speed.

water channeled to each. All infrastructure, roads, and sugarcane transport to the factory was to be provided by Hippo Valley Estates. We had our "bums in the butter" as Rhodesians called such comfort. Financially, the figures provided by government and the sugar company all looked good, and I planned accordingly. In looking over the list of selected farmers, I could see that the drawing had not been random. All had sugar farming backgrounds (the "winner "of the neighboring farm was a chartered accountant from Hippo Valley Estates) or heralded from wealthy farm families, except me (but even I came from a family steeped in dam building and irrigation).

We Hippo Valley settlers, as we were called, banded together, routinely comparing costs and helping one another as much as we could. I soon learned that I was the least experienced and the only one who had to borrow all the necessary capital. Nevertheless, all trended positively, and I focused on hiring the right people. My godfather Eustace's son, John Poles, immediately came to mind. John loved farming and was a close friend. I persuaded John, who had been working for a farmer in Northern Rhodesia since school, to join me. We agreed that we would develop the farm together, clearing the raw African bush and planting our first sugarcane. Then John would live on the farm and manage it while I pursued my scientific work and PhD dream. John moved to Southern Rhodesia with his wife and two children (who stayed with friends near Salisbury while John and I developed the farm; my family—Shirley, Jim, and Megan—also stayed in Salisbury).

Surveying our newly acquired land, John and I realized we had a tremendous amount of work to do. A forest of forty-foot mopane trees had to be cleared—including stumps and major roots—before our rubber-wheeled tractor could plough the land. I had drawn the last farm on the planned water canal, and its water was delivered at the farm's lowest point. Where most farmers with water delivered high on the land could inexpensively flood irrigate by gravity, we would have to build a large holding dam and pump all the water uphill from there to overhead irrigation sprays. Then sugarcane had to be planted, and a diesel engine purchased and set up to pump every gallon of water onto the cane day and night throughout the year.

A close look at the costs—and our limited capital—showed that we could not afford to engage contractors with expensive metal-tracked bulldozers to build our dam, clear the trees, and do the initial ploughing of the land. Instead, we bought two old bulldozers and a twenty-year-old Ford tractor at an estate sale on a nearby ranch. The equipment had been out of operation for years, stored in a dusty farm shed and covered in chicken and bat shit. Though it was thirty miles from our farm, it was close to Buffalo Range, the ranch owned by George Style (whom I had helped begin game ranching). George let John and me camp on his ranch while we overhauled our machinery before moving it to the farm. It was winter and the temperatures were freezing; every morning we'd awake at our camp, bundle up, and ride a motorbike over to the shed. Once there, we'd start a fire to warm our tools and then proceed to repair the three old machines. When we finally had everything working, we drove each machine the thirty miles back to the farm, where we then made camp under a large tarpaulin borrowed from the Rhodesian Railways.

My stepmother, Thelma, as an accountant, proved to be an invaluable advisor. She suggested that I leave the accounting to her and taught me how to manage our farm business by cash flow. She explained to me that monitoring the level of cash coming and going was not the same as preparing the profit and loss statements required for tax compliance—the former being a more essential tool of management. Thanks to her, we did this to an extraordinarily high level, and that was to save us.

John and I hired a third man, an African driver, who drove the tractor while we manned the two bulldozers, and soon after we employed a gang of laborers to dig the trenches for the main irrigation pipelines. By day, we worked to clear trees, bush, and stumps. In the evenings, by the light of a paraffin lamp, I did the cash flow projections. It wasn't long before financial worry set in. It was costly to install underground piping and spray irrigation equipment, not to mention the fuel needed to pump the water onto the land. If we planted our full acreage to cane (which the economists had recommended) and then waited eighteen months (as sugarcane required) for our first income, we would go broke!

We shifted our plan: to keep income flowing, we would also plant faster-yielding crops like potatoes, maize, cabbages, and cotton. Over time, we would progressively increase our sugarcane acreage. This, we felt, was our only hope; we certainly had no means to borrow more money and started to feel a lot less like lottery winners.

CHAPTER 19

CUTTING COSTS . . . AND STEREOTYPES

Learning Farm Management

1963–1964

I have no boss boy on my farm—do you mean my manager, Gilbert Ndlovu?

Gilbert Ndlovu, Kapombe Farm's manager, in front of the first workshop we built (1963).

Little did I know how untimely my farming venture would prove to be amid a shifting political climate—a climate that would get so heated that the Federation[1] would be dissolved at the end of 1963, giving immediate independence from Britian to two of the three countries in the Federation: Nyasaland (renamed Malawi) and Northern Rhodesia (renamed Zambia). With there no longer a "Northern" Rhodesia to distinguish from, Southern Rhodesia became simply "Rhodesia" and returned to being a self-governing colony until April 1964, when—via an internal coup—Prime Minister Winston Field was replaced with a new prime minister, Ian Smith, who would declare

[1] The Central African Federation, also known as the Federation of the Rhodesias and Nyasaland, was commonly referred to as the Federation. It consisted of Southern Rhodesia (renamed Rhodesia in 1964, then Zimbabwe in 1980), Northern Rhodesia (renamed Zambia in 1964), and Nyasaland (renamed Malawi in 1964). It was formed in 1953 and dissolved on December 31, 1963.

Rhodesia's independence from Britian unilaterally and incite more racial violence and disparity than ever.[2]

While aware of this political turmoil, I was not wise enough to see the dangers it posed to my sugar farm endeavor. At the end of Kapombe Farm's first year, European beet sugar came back on the market and subsequently world sugar prices collapsed, showing no signs of recovering anytime soon. The farm would not be able to support both John's family and mine.

A first step was to discuss the situation with John. Much to my regret we agreed that he would leave Kapombe Farm to work as a farm manager for a farmer near Gatooma (now Kadoma) in the center of the country. I would in turn carry on the best I could. However, I was becoming increasingly involved in the army (which I elaborate on in Part VIII, Irregular Warfare), and that took me off the farm for weeks at a time. I also desired to spend more time supporting my friends' game ranching efforts. I was going to need assistance on the farm; I needed a reliable manager.

After putting out the word, a local African, Gilbert Ndlovu, who had run his own building business, called on me. His business had failed, and he was desperate for work. Gilbert answered all my questions well and had an attitude I liked. After a good discussion, I hired him (even though he, like me, had no farming experience). Gilbert started with a meager salary (though higher than what I regularly drew from the farm) and the promise of a bonus based on crop earnings. Immediately, I took him under my wing, teaching him all I could about management of time, labor, cash, crops, and maintenance, which I was myself learning on the job.

There was one hitch, and that was combatting existing racist stereotypes. With frequent absences, I needed fellow farmers and local shops to accept that Gilbert had authority as my farm manager. Almost every farm in Rhodesia in those days had a "boss boy," typically a Black African *man* who oversaw the labor. But it was unheard of for a farm to have an African as a manager. I would go into a local shop, and the owner would say, "Oh Allan, by the way your boss boy was in yesterday and ordered so and so, but he didn't have a note from you, so I was hesitant to provide."

I would respond, "I have no boss boy on my farm—do you mean my manager, Gilbert Ndlovu?" Gradually, merchants and neighbors accepted that I had an African as a manager and started referring to him by his name, Gilbert.

Gilbert and I worked well together, employing a motivated team of workers, all of whom we housed in mud and thatch huts, even though I continued living under a tarpaulin cover (my family would join me when I had a proper place for them to stay). The biggest challenge was cutting costs, which forced me to think creatively—perhaps more

[2] I go into further detail regarding the political changes of Rhodesia in Chapter 32 and beyond.

so than in any other period of my life. Cutting costs could not be done by decreasing staff, lowering wages, or expecting the team to work even harder. On the contrary, we needed more laborers—with a spirit to do more.

The work patterns of our men varied greatly—some were natural workers, some loafers, some clowns, and some barrack room lawyers, as we called the troublemakers. We aimed for an ideal team of all natural workers—workers who were self-driven and required minimal supervision. Gilbert and I gradually began to select and dismiss men. To determine who stayed, we would first give the whole gang a difficult task, then Gilbert and I would go to town, leaving the men unsupervised. What the men did not know is that "going to town" for me meant climbing the densely bushed hill behind my camp, where I would then perch, observing the crew through binoculars. I would take note of those who kept working without supervision; those who sat in the shade smoking were dismissed at the day's end. When the dismissed asked why, Gilbert and I would simply say, "You are not the sort of person we need." We would then have little trouble taking on other men, as many in those parts wanted work. This daily selection process went on until we had all natural workers, with a few natural clowns, as I called them, who made light of hardship and kept the men laughing throughout the day. Each farm around Kapombe grew one or two crops (generally sugarcane and maize or cotton) with a labor force averaging eighty men plus contractors to clear the mopane forest, build the dam, and prepare the land for planting. Kapombe Farm, on the other hand, put in five crops, with a labor force of forty men and no contractors; we cleared the trees and built our dam ourselves.

Sometimes, it was touch and go. Our cotton crop absolutely had to provide the yield I estimated and come in on time or we would not survive, and we watched in dismay as weeds engulfed the tender plants. Our men could not keep pace weeding manually and to use the tractor would do too much damage as the weeds grew as high as the cotton. Nor could we afford spraying herbicide. In desperation, I took to my hill to puff my pipe and think. When I came down, I went straight to Gilbert.

"I've got it! We can make it work!"

We divided the cotton field into four equal blocks—which was easy to do with the irrigation lines evenly spaced as they were—and Gilbert nominated four of our best men to support the task. To each of them I explained, "From today until we pick the cotton, you oversee weeding. Each of you have one block of cotton. We will sign off on your work every day, whether you work or not—it is entirely up to you. All we ask is that you keep your block as free of weeds as possible by simply pulling them out."

The selected men thought I was joking. "How on earth can four men be expected to do what forty men could not?" they inquired. I urged them each to do the best they could to keep their block clear of weeds—there were no rules. I then added a twist, "Every

Saturday morning, Gilbert and I will judge each of your blocks. Whoever has the fewest weeds in their block will get an extra day of pay and a gallon of beer."

The four men each did their best, and the following Saturday—in front of all the men, their wives, and children—we announced a winner (who had done slightly better than his three competitors). In front of everyone, we made a fuss of the man, congratulating him with the promised extra day of pay and a gallon tin of beer. Much clapping, singing, and laughter ensued. I would then focus more attention on the winner by challenging him to drink the gallon of beer without taking it from his lips. If he did so, I'd give him another gallon. He would inevitably fail the second time, spilling beer all over his chest and joining the whole team in laughter.

On Sunday the four men were back at it again, weeding their blocks, but this time two of them approached us with concern. They wanted a ruling from me, because, they said, it was unfair that the other two men had their wives and children helping them. I reminded them that there were no rules. They could ask anyone to help them, take a day off, do as they like—it would not change the fact that the winner would be the one whose block had the fewest weeds come Saturday.

By Monday there were many people in each block of cotton—the wives and children of all four men had joined, singing, laughing, and pulling weeds. By Tuesday afternoon I asked Gilbert, "What's going on?" There were more than a hundred jubilant singing people pulling weeds, many of whom I did not know at all and clearly more people than we had on the farm. Gilbert explained that the laborers on nearby farms had heard of the weekly competition and were placing bets. They had now figured out that they could also bring along their wives and children to help the "horse they had backed"! Every week until our cotton was picked, we gave the prize, though we literally had to start tossing a coin because not a single weed could be found! We had saved the farm—and everyone's jobs—with an incentivized competition costing one day of extra pay and a gallon or two of beer a week at one shilling and sixpence a gallon.[3]

In addition to creatively thinking about labor, we schemed ways to cut our machinery maintenance costs to amazingly low levels—so much so that we never needed more than the two bulldozers and twenty-year-old tractor. In the mutual help sessions with the other lottery farmers, we learned that our machinery costs were on average half the machinery costs of theirs. The key to this cost cutting was that in doing my own repairs, I was able to get to the root cause of every breakdown, concluding that breakdowns were almost always started by something loose or dry and unlubricated. Whereas my neighbors treated their workshop as the place for both maintenance and repairs, I changed that on Kapombe Farm. Maintenance became the driver's or machine operator's responsibility;

[3] The beer we awarded was the traditional opaque beer brewed with a variety of locally grown sorghum.

the workshop was solely reserved for repairs. To incentivize the drivers, I laid them off as employed drivers and rehired them as contract drivers running their own businesses with the potential of higher earnings. Now, maintenance oversights came with penalties, and maintenance checks could occur at any time. It was amazing to see, with the focus on daily maintenance, how much our repair costs dropped. They were even lower than the repair costs of those of neighbors boasting brand-new machines.

CHAPTER 20

SAND AND BASALT

Understanding Soil

1964–1965

I had never been taught—nor gleaned from a textbook—that there are four processes so interlinked, they might as well be one.

Kapombe Farm, fully developed, planted with sugarcane, and under overhead spray irrigation (1965).

This brief spell in my life of actively growing crops under overhead irrigation on two vastly different soils—sand and basalt—was to teach me much as an ecologist, and my learnings became central to solving the riddle of desertification. Where I had previously learned about basic ecosystem functions, I had never been taught—nor gleaned from a textbook—that there were four processes so interlinked, they might as well be one. This I learned on Kapombe Farm. These four processes are:

1. *the water cycle*—the water that cycles between earth and the atmosphere;
2. *the mineral cycle*—the mineral nutrients that cycle constantly among soil, plants, and the atmosphere;
3. *energy flow*—the energy that flows from the sun through plants and then into everything that consumes them; and
4. *community dynamics*—the communities of diverse organisms above and below ground that are constantly changing, including in response to human actions.

These four processes can only be managed successfully when viewed as one. The slightest change in any one of them changes all of them.

The link between the water cycle and mineral cycle became clear, for example, when extreme mineral deficiencies showed up in patches of Kapombe Farm's sugarcane crop. Agricultural extension and research experts advised that we "fix" this through fertilizer applications, however I wasn't convinced this was the best approach. I reasoned the minerals had to be there, as the soil where the deficiencies were showing in my cane was an incredibly mineral-rich basalt of volcanic origin. I concluded that the minerals must be "locked up" or unavailable to the plants in some areas, but I was not sure why.

My father, with a life in irrigation, introduced me to an Italian man whom he was confident could help me. The man, who had no academic training but years of field experience, arrived and immediately asked to use my soil auger. When he saw our four-foot-long auger, he scoffed at the length, saying it was too short. We got to work welding on another rod, which extended the auger to just over six feet long. The man then drilled down into the mineral-deficient cane areas, evident by the streaks of different colors in their leaves. Once the auger reached six feet, he simply pulled it up and down. "Listen," he said. I could hear a sucking sound accompanying the auger's movements. "That!" he exclaimed, "is your problem."

The problem? We were overwatering these areas, making it impossible for the deeper cane roots—with insufficient aeration—to access the full spectrum of nutrients in the soil. We cured this by simply changing our management of the spray lines and sprinklers. We stopped applying water to these areas until all the plants had wilted *and* then started regrowing—once their roots had drained the waterlogged layers of soil. Thereafter, in the areas prone to waterlogging, we pulled out sprinklers along the spray line after a limited time. The "extreme mineral deficiencies" problem was solved.

Soon after this lesson, I learned another that demonstrated how improving just one process improves all four. We had noted how the sugarcane (not to mention the weeds!) was growing slower where it had been planted in furrows (per the advice of experts) than the weeds which grew profusely on the ridges. With this observation in mind, I decided to conduct a little experiment using cabbages (they'd grow faster than cane and hence help me draw more immediate conclusions, not to mention I could also sell them at the end of the trial). The cabbage seed packet said that the cabbages should grow to a weight of approximately two pounds. With this in mind, we prepared an area of land using our cane ridger. We carefully made the furrows and ridges per the experts' advice. This time, however, we also left flat areas between the ridges and furrows. We planted the cabbage seeds in all three places: the furrows, the ridges, and the flats. A couple of months later, we harvested the cabbages from all three places and weighed them. On the flats, the cabbages grew to the expected two pounds each, as advertised on the seed packet. In the furrows, the cabbages averaged a disappointing one pound each. On the ridges, however, the cabbages averaged an unbelievable eight pounds! How could this be?

On the ridges we had not changed the water given, the nutrients in the soil, nor the amount of sunlight. We had only changed the amount of time plant roots were in soil with a good balance of air and water. The plants on the ridges, which rapidly drained off excess water to the hollows, did not spend valuable growing hours with their roots in excessive water and minimal aeration. Consequently, between irrigation applications those cabbages were growing at a maximum rate with every hour of sunlight. In the furrows (where I'd been advised to plant), after each irrigation the roots were in soil with too much water and too little air for several days; the cabbages could only grow slowly until the air/water balance improved.

What a lesson on the importance of soil aeration (air and water in balance) to an effective water cycle. Good aeration also enhanced the mineral cycle, making more minerals available to the plants, and enabled plants to grow better and thus trap more sunlight. Good aeration, in fact, improved the health of the whole community (plants, soils, and soil life). Of course, even before harvesting the cabbages for sale, we changed to planting all our cane on top of the ridges—exactly where researchers advised us not to![1]

---

[1] I would later learn that, 4,000 years ago, farmers in Mexico had solved their waterlogging problem in wetlands by growing their crops in raised beds known as chinampas.

CHAPTER 21

WHAT MATTERS MOST

Schistosomes, Meetings, and Moving On

1965

We farmers, each wanting to have our say and unable as a united group to focus on what mattered, too easily became clay in the hands of skilled, self-serving officials.

My children Megan and Jim with baby porcupines they'd caught (I made them return the babies to their mother).

While I was developing Kapombe Farm, Shirley had remained in Salisbury, holding down a job as a botanist at the National Herbarium. In late 1964—once I had a better place to live than under the tarpaulin cover I had called home—Shirley, Jim, and Megan joined me. We sold our Salisbury house, and everyone moved into the brick workshop (which would serve as our house until a more fitting one could be built). We were once again together as a family. Jim and Megan were growing up fast, and we were thankful to be surrounded by other young farming couples with children, creating a pleasant social life despite the frontier-like conditions.

All the water for my family and the families of our farm labor had to be drawn from the irrigation canal, which unfortunately was full of schistosomes—the parasite that causes the deadly bilharzia[1] disease. On advice from health and bilharzia experts, our neighbors were treating their water with copper sulphate, which killed the parasites in the water before they could infect a human. I knew of no research that had studied the effects on children of drinking water laced with copper sulphate, so I sought another solution. Friends in the Bilharzia Research Laboratory in Salisbury allowed me access to their library of papers on the life cycle of schistosomes. I quickly went to work doing my own research. I learned that schistosomes leave their host as eggs in feces or urine. When these eggs enter water, they hatch and penetrate snails. There, they asexually reproduce and reenter the water as cercariae (small fork-tailed worms) that must find a host within twelve hours or die.

Armed with this information, on Kapombe Farm I realized we had no need to lace our water with chemicals. We constructed two water tanks connected by a pipe with a tap in the middle. Our cook was put in charge of the water flowing from these tanks; he received strict instructions to only open the tap between the tanks every twenty-four hours. We would then draw our drinking water from the second tank, confident that more than twelve hours had passed in any cercariae's attempt to live. Jim grew to six feet, four inches—well above my six-foot frame—so I guess children thrived on dead schistosomes. Neither my family nor my laborers got bilharzia.

Kapombe Farm soon became a place for our extended family to get together on various holidays. After an Easter gathering one year, my father, his wife Thelma, and my sister Ann were returning to Salisbury when their car, which Ann was driving, slid off the sharp edge of the tar-strip road onto the gravel and overturned. I received a call from the police and hastily drove to Chipinga Hospital. When I arrived, the hospital had already transferred Ann—who the doctors felt had sustained the worst of the injuries—by air (courtesy of a volunteer's light plane) to Salisbury General Hospital. In lieu of there being no ambulance, I drove my father and Thelma the five hours to Salisbury. Tragically, the doctors had underestimated the seriousness of Thelma's condition, and she died shortly after our arrival. Ann, on the other hand, lay in a coma for days. I sat by her bedside, willing life into her hour after hour. (I am not religious; however, having gone through this, I have an appreciation for the comfort of religion when faced with trauma.) Eventually Ann recovered from what was diagnosed as severe brain bruising, yet to those of us who had known her before, she was never the same. Her personality changed; from thereon she was easily irritated and much less spontaneous and fun-loving.

Alongside this family tragedy, my optimism for farming started to plummet. Despite all our efforts, the cash flow that I projected out five years showed a steadily deteriorating picture. My neighbors assured me, "Allan, this is just a dip; government will not let this

[1] Bilharzia, also known as schistosomiasis or snail fever, is a disease caused by parasitic flatworms that live in freshwater snails. Symptoms include abdominal pain, diarrhea, blood in the stool and urine, itchy skin, fever, chills, cough, headache, joint pain, and sometimes paralysis.

scheme fail!" Rod Dacomb, my neighbor who was also a chartered accountant, told me I was worrying too much.

Finally, I sought his help. "Rod, may I show you all my projections?" I asked. "Please study them and tell me where I'm wrong, given all your accounting and sugar business expertise." A few days later, he sheepishly responded, "Allan you're right, and we're in trouble!" All thirteen farms were in financial jeopardy to some degree, however the six farms—including mine—that were unable to flood irrigate faced the worst financial outlook (we six had the additional expense of pumping irrigation water day and night). The chances of survival were indeed bleak.

Because the settlements in Hippo Valley were a government scheme, all thirteen farm owners decided to initiate a series of meetings with government. I was about to experience what it felt like to be a member of the public dealing with civil servants. Before each meeting, we farmers would agree on and carefully prepare our list of concerns and priorities. However, at every meeting—in only a matter of minutes—either government officials or a member of our group would go off topic. This cleared the way for civil servants to avoid accepting responsibility for poorly surveyed farms and irrigation canals. We farmers, each wanting to have our say and unable as a united group to focus only on what mattered, too easily became clay in the hands of skilled, self-serving officials. Finally, fed up, I walked out of a meeting.

When the meeting ended, my fellow farmers rushed out of it jubilant, exclaiming, "Allan, you should've stayed! We made great progress!"

I in turn asked, "What progress?"

All they described simply added up to the need for further meetings!

I decided to hold my own meeting at Kapombe Farm, inviting the owners of the five farms that, like mine, were in the direst of straits. Each farmer duly arrived; I promptly served tea and handed out pieces of paper containing the three points we had always agreed *had* to be settled. I waited patiently as the farmers read the points.

"Are these the points we have to negotiate with government?" I asked. All agreed. "Then why the hell will you not stick to them? In every meeting, despite our firm agreement to stick to the main points, you fellows always deviate and allow the officials to take you on a detour!"

Keeping in mind that my family's future was at stake, I continued, "I'm determined to focus on what matters; we must form a united front and meet with the prime minister. Who will join me?" A ripple of protest erupted through the group, the surprised farmers claiming a meeting with the prime minister was impossible. I simply repeated, "Who will come with me?" All agreed that if I could get that meeting—which I assured them I would—they would accompany me.

I then said there was one more issue we needed to settle and agree upon now, not later. Because—based on my experience in previous meetings—I had little confidence that any of them would stick to their guns, we had to have one spokesperson only, and that person would be me. Again, this was met with protest, several farmers accusing me of being dictatorial. Again, after much discussion, everyone calmed down and agreed that I would be the one and only spokesperson. In turn, I committed to securing the meeting with the prime minister, representing all the farmers by sticking to our main points, and allowing no deviations. I went so far as to say that in any meeting, if one of them opened his mouth, I would walk out. (I had to be that strict if we were to have any chance of success. Unfortunately, as I have learned many times since, it is truly difficult for decent people to concentrate on what matters most, even, as in this case, if it means saving their own families.)

All six of us then drove to the headquarters of the Commercial Farmers Union (CFU) in Salisbury and asked the staff to arrange a meeting for us with Prime Minister Ian Smith. Smith refused to see us, relaying instead that his minister of agriculture, George Rudland, would meet with us. I agreed to the meeting, reminding the rest of the farmers that they were to keep silent. We filed into the meeting room, seating ourselves at one end of a long table headed by Minister Rudland and lined with officials.

Rudland—a well-known character and a wealthy, ruthless, domineering businessman—welcomed us, stating he was ready to discuss our concerns. I replied, "We are not in this meeting to discuss our concerns and have not, in fact, requested this meeting."

Confused, Rudland asked, "What then are you here for?"

I replied, "We are here to determine a time and place at which to meet with the prime minister."

Rudland became angry, accusing us of wasting his and his officials' time, shouting, "The prime minister will not be meeting with you!"

I calmly replied, "Mr. Rudland, don't get angry with us. You called this meeting—we didn't. We have come to see the prime minister; we have nothing to discuss with your officials—nor you—that we have not already tried to discuss many times."

Rudland glared at the others on my team, asking, "Haven't any of you anything to say?" I held my breath, knowing this was the crucial make-or-break moment. Thankfully the rest of the farmers stood firm.

"No, Savory is our spokesperson."

We were all promptly dismissed, only to return to CFU headquarters the next day, again requesting a meeting with Prime Minister Smith. This time, we were again told that he would not meet with us but was handing us off to minister of land and water development Phillip van Heerden. Again, we filed into a meeting room and stood united, clarifying that we had only one item to discuss—the time and place to meet the prime minister.

Undeterred, on the morning of the third day we marched back to CFU headquarters, once more requesting a meeting with the prime minister. This time Prime Minister Smith agreed to meet with us—but not until 4 p.m. I knew nothing about Smith other than he was a popular and slick politician. On being shown into his office, his first words were, "Which one of you is Savory?" I said I was, and I deliberately sat next to him rather than across the table.

Smith did not waste time getting right to the point, "It's no use to meet with me; this is not a government scheme. Rather, your concerns are Hippo Valley Estates' issues, and you'll need to meet with them."

Having anticipated this move from Smith, I respectfully disagreed and opened my file, showing him paper after paper of his government's promotion of a government settler scheme—the advertisements, the application forms, and so on. Seeing the mass of evidence in my hands, Smith wisely chose not to argue. He simply sat back, sighing. "All right, you've got me! It is a government scheme, but the truth is we haven't the money to put it right."

I thanked Smith for the first honest reply we had received from government in six months, then urged that we find a constructive way forward. Within thirty minutes we had a solution. "Government," Smith said, "will arrange for Hippo Valley Estates to buy out those of you in the worst position. You will be given priority for resettlement elsewhere should opportunity arise."

We thanked Smith and walked out. Immediately I called my father, informing him of our meeting and that we would be selling Kapombe Farm. He was reluctant to do so because we would lose so much financially, but I reminded him that I had the final say, and he honored our earlier agreement. I promptly sold the farm to Hippo Valley Estates and applied for resettlement. From my time in the army (covered in Part VIII, Irregular Warfare) I had already scouted out the wildest places in the country, setting my sights on a large block of land bordering Botswana and embracing Kazuma Pan.

The "pan" was a large depression where water collected and thus was an attraction for many game animals. Up until 1949, Kazuma Pan (hereon simply referred to as "Kazuma") and the land surrounding it had been an undeveloped national park lying between two forest reserves. It was so remote that no one ever visited. Believing the land to be "useless for any purpose" the government had that year changed its status to state land and leased it to the Rhodesian Hunters' Association, whose members regularly hunted it. The government agreed to lease me 37,000 acres of this land for twenty years, with an option to purchase it after two years at two shillings and sixpence an acre. This meant an initial lease fee of about $470 per month[2]—making Kazuma reasonable compensation considering my heavy losses on Kapombe Farm.

Shutting down Kapombe Farm—after selling off the equipment and paying the staff for the last time—was painful but the right thing to do; it left me broke but debt-free

[2] This is equivalent to 3,600 (US dollars) in 2025.

(except for what I owed to my father). On our final day at the farm, I hitched a trailer to our old but ever-dependable tractor and loaded it up with everything that didn't fit in my Land Rover.

As we were about to embark on our 150-mile journey (at tractor speed) to our new home, my trusty manager Gilbert came up and loaded his bedroll and pots on the back of the Land Rover. I assumed he was hitching a ride and asked, "Where will you be heading next, Gilbert?"

To my surprise, Gilbert replied, "I'm coming with you."

I told him I was broke, "I have no money to pay you; I'm going to Liebig's Ranch (where my friends had leased over a million acres for game harvesting) and will be camping under a tree on the Umzingwani River with Shirley and the children until my Kazuma lease is finalized."

Gilbert chimed in, "I will camp under the next tree." And with that he climbed onboard, adding that he knew I would pay him someday when I could.

Gilbert did not leave me for the rest of his life. Little did I know at the time, but Gilbert's presence would save me from later "terrorist" attacks during the escalation of Rhodesia's civil war, when my family and I would be living in the bush—far more remote and isolated than any farm or ranch in the country.

PART V GAME RANCHING

CHAPTER 22

WHAT'S IN YOUR TOOLBOX?

Harvesting Game Humanely with Resourceful Hunting Teams

1965–1967

Our mission rested on the hypothesis that, if the harvesting of game proved to be of monetary value, landowners would in turn protect their wildlife as a valued resource.

Peter Johnstone (*right*) and a friend sitting on Peter's Land Rover at my old tsetse department camp at Dotts Drift in 1959, a few years before he joined the WUS team.

As it turned out, finalizing my lease (and subsequent purchase) of Kazuma would take many months, and it would be more than a year before I could move there with my family. I had outwitted several senior civil servants in obtaining this lease as settlement for Kapombe Farm, and they retaliated by putting every bureaucratic obstacle imaginable in my way. Because the game department profited from leasing Kazuma to hunters, it bitterly opposed my plans to establish privately funded wildlife management research there. Meanwhile, I had to find a means to continue supporting my family.

On moving down to Liebig's Ranch,[1] I became fully involved in the game ranching business. While in the Southern Rhodesian Game Department, I had helped Archie Mossman, Peter Johnstone, and Robbie Robertson (an old friend with good business experience) establish Wildlife Utilization Services Pvt. Ltd. (WUS), a company on whose board I had been serving since leaving the department. WUS was leasing the hunting on Liebig's Ranch. We intended to show that we could manage the wildlife in a way that would not only prevent desertification but reverse it *while also profitably providing food for people.* Our revenue model was strictly focused on the meat and hides harvested from wildlife (it did not include income from sport hunting or photographic safaris). Our mission rested on the hypothesis that, if game ranching proved to be of monetary value, landowners would in turn protect their wildlife as a valued resource.

At the time, my research as well as that of Archie Mossman and Ray Dasmann confirmed the accepted scientific view that too many animals led to environmental damage (hence the need to cull elephants in the Zambezi Valley). WUS took it on faith that, if we could get the numbers right and harvest game humanely and efficiently, revenue would come in higher and expenses lower than ranching models based on the production of cattle, sheep, horses, donkeys, and goats. Subsequently the land would improve because native animals were, well, native.

Of course, there was much we hadn't thought about and overlooked. We had reason to think that tsetse flies and ticks would not be existential problems; we did not reflect much on why and how humans had domesticated wild animals several thousand years prior or exactly how this domestication differed now; and we hadn't contemplated why everyone, including us, felt land only deteriorated under domestic grazers but not under wild grazers (as many people still believe today).

Nonetheless, in 1965 I marched ahead with the mission of WUS and, in now camping and living on Liebig's Ranch, became chairman of the WUS board in addition to being one of the hunters (the latter a dire need as we couldn't afford to hire many hunters). Archie and I continued our focus on research and game assessments, while Peter and Robbie oversaw the hunting and marketing.

Because of a combination of distances, lack of capital, and intense opposition from Rhodesian cattlemen who resented any business that competed with beef, WUS marketed game skins and dried meat in just two forms: game biltong (from the best cuts of large muscles brined and shade-cured), which we sold by the pound, and ration meat (all other meat, salted and sun-dried), which we sold by the 200-pound sack.

[1] The Liebig Ranch, commonly referred to as Liebig's Ranch, was a 1.25-million-acre ranch owned by the London-based Liebig's Extract of Meat Company. Arguably the largest cattle ranching company in the world, it owned ranches in Argentina, Paraguay, Uruguay, South West Africa (now Namibia), Australia, and Rhodesia. These ranches supplied the meat for the company's famous Oxo extract and later its Fray Bentos corned beef. Wildlife Utilization Services Pvt. Ltd. (WUS) would also lease the hunting on the forestry commission's 357,000 acres at Gwaai Forest—a teak forest between Bulawayo and Victoria Falls.

Our butchery was built entirely of gum (eucalyptus) poles and wire netting. The poles and netting held hessian cloth (burlap) coated with cement to make thin weatherproof walls that lasted for years. Though rudimentary in its construction, the butchery was large enough to handle animals of all sizes, including elephants. My tractor and trailer proved useful in hauling all but the largest bull elephants into the cutting shed. In the case of elephants, we first had to hoist up the carcass with block and tackle suspended from three long metal sheer legs. Once the carcass was high enough to slide the trailer under it, we did so and brought it in.

Based on our earlier game ranching experience on Doddieburn Ranch, Peter and I organized hunting teams comprised of one hunter who drove and four men as trackers and loaders. Each team had one stripped-down, secondhand, short-wheel-base diesel Land Rover; a two-wheel trailer; and a basic toolbox containing the essential tools needed in case of a breakdown. We had no radios, maps, or phones. Each day four or five hunting teams dispersed over more than a million acres, operating hundreds of miles from one another and seeking game of the species we were harvesting. We were highly adept at looking after ourselves, harvesting the necessary game, and getting the animals to the butchery in acceptable condition; spoiled meat was never excusable, not even in the hottest season.

I could write an entire book on the feats accomplished by our incredibly self-reliant and creative hunting teams. The largest load ever brought back was sixteen zebra by five men in a short-wheel-base Land Rover pulling a compact two-wheel trailer. The men had packed the zebra in closely: three on the bonnet (hood), one across the driver and passenger seats, three in the very short back of the Land Rover along with the men, and the rest piled on the trailer. The hunter who drove sat on top of the zebra across his seat, leaning down to steer while a tracker—sitting as well on the carcass but on the passenger side—used a long, forked stick to push either the clutch or brake, pending the hunter's instruction.

No matter the mishaps, it was the hunter's job to get his team and any game they shot back to the butchery somehow, because no one knew where they were nor could come to their rescue. Subsequently the hunters became amazing bush mechanics. One hunter arrived at the butchery very late one night but with his harvest still usable. His delay was due to a back wheel bearing on his Land Rover disintegrating entirely. Realizing his dilemma, he had built a fire, removed several bullets from his rifle cartridges, placed them in a pot, and melted the lead out of them. He had then poured the molten lead into the remains of the bearing housing and smoothed it off with his knife to make a lead-to-lead bearing that got him home!

Because both Peter and I were also hunting, we developed ever-improving techniques that allowed us to kill more humanely and with minimal wounding. Peter kept meticulous records; on our busiest day, the two of us harvested 100 animals, killing them humanely using only 1.03 rounds per animal. No animals were wounded and lost.

I could not say the same for myself. One day, while running down a herd of eland (which are easy to run down because, after their first fast gallop, they never break into more than a trot), I twisted my ankle. It gave way with a loud crack. I continued in pain, but it severely worsened overnight. After a few days' rest, I stubbornly resumed hunting, but that was a mistake. My ankle ended up requiring a rather complicated operation involving my Achilles tendon. A master surgeon, Standish White, did his best but to this day my right ankle and foot can only move up and down, not sideways.

On Liebig's Ranch, Gilbert and I were once again tasked with cost cutting, relying on much of what we had learned during our Kapombe Farm days. Two of our most notable cost-cutting methods were focused on the tools and the hunters.

The loss of tools was a reoccurring problem with each of the hunting teams. We stopped this entirely by simply having hunting teams show a full toolbox as they came for their pay. If a tool was missing, they received no pay until it was found or replaced. Tool loss dropped to zero.

The wages of hunting teams were our biggest expense, but the hunting teams were also our sole source of income. By hunting with them, I noted how much the leader of each team varied in skill, fitness, and determination. To weed out the poor performers, Gilbert and I decided on a policy resembling that of our worker selection process at Kapombe Farm. We paid the Black African trackers and loaders in each team a fixed daily rate, regardless of results. However, we laid off and then reemployed the European (White) hunters leading each team as contractors with no salary or wages; they were paid entirely on results. The amount paid was based on the weight of each animal once skinned, with bruised and damaged meat trimmed off. By tying the hunters' pay to their harvest, we saw an immediate overall financial improvement. The best hunters earned far more than they had on any fixed wage, and those who could not make enough to support their families left. By rewarding our most effective and efficient hunters, they kept our business going, even during the toughest of times.

CHAPTER 23

HOME, SWEET KAZUMA

My Family's Best Years

1967–1968

The four of us would sleep in a row on the ground, jokingly telling Meg that she, as the smallest, was our lion bait.

(*left*) With Jim, Meg, and Shirley huddled around the campfire at Kazuma on a cold winter morning. (*right*) My horse, Bullet (with zebra skins in a sack on his back)—there wasn't a better horse.

In 1967, after nearly two years of struggle with the Southern Rhodesian Game Department, which did all it could to block my purchase of Kazuma (even though government had agreed to it), I managed to move around them. I'd had to phone the Secretary of Lands every week to keep things moving (always asking if I needed to talk to the prime minister again) until the government finally granted me the agreed lease with option to purchase. (My plan was to set up a nonprofit company intended mainly for wildlife research, naming it Kazuma Range and Research Station. I invited the National Parks Department to have two seats on the board, but they declined.)

Unfortunately, the Kazuma I took over was not in good condition. The Rhodesian Hunters' Association had burned the grass each year to make their sport easier (and possibly less sporting), leaving the land in a very sorry state. But at last, I had my own land to manage. Accordingly, I made plans to move to Kazuma, once more with next to no money. My position as chairman of WUS (while also hunting for WUS) paid little, as did consulting with other game ranchers. I was spending increasing periods in the army, for which I was paid; however, because I was a "territorial," the pay was well below that of regular army officers of the same rank.[1]

At the first opportunity, I moved my family up to Kazuma, along with Gilbert and four men (the most I could afford to pay). We selected a grove of large, shady mfuti trees[2] on the edge of a Kalahari sand forest ridge and mopane veld as our new home site. We built the first hut—of bush poles and thatch—for Gilbert as he stayed there even when I had to be away. Next, we built huts for the four men I'd hired. Each hut was made of mopane poles and ant heap clay taken from the land. The roofs were thatched with grass. The only materials we had to purchase were twine (for tying the grass on the roofs), wire (for tying poles together), and a few nails. We used old curtains for doors. To keep costs low and Kazuma as wild as possible (as I intended it to become a wildlife research station and ultimately a national park), we made minimal road tracks. For transport, we mainly used horses or my faithful tractor from Kapombe Farm (I had stripped the tractor of all costly-to-maintain electrical components and now started it by pulling a rawhide rope wound around the power offtake).

Meanwhile, Shirley, Jim, Meg, and I camped under the trees, constructing our new home around the campfire where we gathered, ate, and slept. In many ways those Kazuma years were to be our best as a family, although our hardest. Until we had built the first huts, the four of us would sleep in a row on the ground by our campfire, jokingly telling Meg that she, as the smallest, was our lion bait. Each night we took turns reading by the light of a paraffin lamp. The first book we read was *Jock of the Bushveld*,[3] Sir Percy FitzPatrick's wonderful story about his life as a young man who, in the company of his fierce and faithful dog, Jock, transported goods in a wagon from the coast in Mozambique to the interior of South Africa through what is largely Kruger National Park today. Like so many children in southern Africa, I grew up with FitzPatrick's tales. Now, with us huddled by the fire, his adventures became ever more real, accompanied by the live soundtrack of roaring lions or trumpeting elephants in the distance.

Within a couple of months our home was completed. Jim and Meg each had their own small hut alongside Shirley's and my bedroom hut, and we'd also built a sitting room hut. Each hut was free standing for ease of building but mainly to reduce the possibility of losing our entire home to fire. When Jim and Meg grew old enough to attend school, we rented a cheap house on the far north edge of Bulawayo. This second home had also become necessary because by then I was in Parliament[4] and was also frequently called up by the army for months at a time and would not leave Shirley alone in the bush.

---

[1] I expand on my army endeavors in Part VIII, Irregular Warfare.

[2] Mfuti trees *(Brachystegia boehmei)* are indigenous to eastern and southern Africa; they are known by various names, including "Prince of Wales' Feathers" in English.

[3] Today one can buy only a politically sanitized version of the 1907 book *Jock of the Bushveld*; FitzPatrick's story has lost much of the character of the Africa of those days, which, for all its sorrows, had its glories, too.

[4] I expand on my time in Parliament in Parts IX (Parliament) and X (The Opposition).

To let the game on Kazuma settle down after years of intense hunting, I chose to harvest only enough to feed my family and staff. The government would not allow tax breaks for money I spent for research on game or land management *unless* I showed income from the property. Determined to make ends meet—and with no funding for any nonprofit work—I spent each Christmas holiday (when the children had six weeks off from school) hunting zebra and selling their skins. Though I killed about fifty zebra a year to show a passable income from the land, I could not transport the meat from the remote bush without adding considerable cost and labor, so I left it to the lions and hyenas. My men had said they would not eat zebra for reasons tied to their clan totem, but when they saw me leaving meat in the bush, they changed their minds and ate zebra without any adverse effects.

When Meg turned nine, I presented her with her first horse, which she loved dearly and quickly learned to ride. One day, when we were out riding together, we saw a zebra stallion on his own. I dismounted to shoot him, holding the reins over one arm, but as I fired my rifle, my horse (coincidentally named Bullet) jerked his head, and I managed to only wound the zebra, which ran off. At this point it was getting late in the day and—knowing I had to leave early next morning for the army—I instructed Meg to ride home, which was about three miles through the bush. With Meg on her way, I turned to ride the zebra stallion down and kill it before dark.

Riding down a zebra on horseback is far more difficult than on foot. On foot, as I noted in Chapter 16, I had learned to run at an angle on one side of the zebra herd, pressuring the zebra to circle to keep you in their sight. However, on horseback, zebra run straight away from a horse, not caring to keep horses in their sight. This ride turned out to be a rough one indeed. Eventually, after crashing through bush and leaping over dongas,[5] Bullet and I were able to get alongside the stallion and I killed it from my saddle using my Colt 45 (which I carried strapped to me, in case I was ever pulled off my horse by a lion). To my surprise, seconds later, Meg arrived on her horse. She had decided to follow me, not wanting to miss out on the chase. I could not believe my little girl had ridden at full gallop over such rough country; I had to give her some hell for disobeying but could not hide my pride in her.

We were remarkably lucky when it came to dealing with predators around our horses, especially since we had intentionally chosen not to build fencing because it would interfere with wildlife movement. Our horses wandered wherever they liked during the day. In the evening, we called them into pole pens and gave each a tin of mielies (corn kernels), groomed them, and checked their feet. There was no farrier within 300 miles, so our horses were never shod, but we kept their hooves filed, mimicking the rounded edges and angle of zebra hooves.

[5] Dongas are deep, sharp-sided gullies.

Once at about midnight, one of my staff woke me to tell me that Bullet had jumped out of his pen. I pulled on my running shorts and—guided by moonlight—quickly found Bullet, haltered him, and led him back to his pen. Bullet however refused to enter. I attempted to reassure him, shining my torch around his pen to show him nothing was there, but it made no difference. He would not budge. Finally, I got the man who had woken me to hold Bullet's halter while I scouted the premises, carefully circling the outside of Bullet's pen. Much to my surprise I flushed a huge male lion lying up against the pen, hidden by a bush, waiting for me to put Bullet back. Each time I had shone my torch, the lion had evidently closed his eyes, so I did not see him. As soon as the lion ran off, Bullet easily entered and settled down. One afternoon years later, lions did kill Bullet, but he was the only horse we ever lost to them. One of our horses came in one night with great claw marks down both sides of his rump; we could only guess he had kicked that lion into space!

Jim, Meg, and I spent many days together in the bush and those days remain some of the most precious to me. Once Jim was old enough, he started hunting with me (one life ambition achieved!). One morning while teaching Jim to hunt, we silently stalked a herd of waterbuck.[6] Suddenly, the animals lay down flat, the bull turning his head sideways and tucking his long horns into a bush. I knew this behavior was not due to our being detected; I whispered to Jim to just sit tight to see what we could learn. After a good fifteen minutes watching the waterbuck remain perfectly still, we heard the faint sound of a forest service tractor way off our land and put the puzzle together: the waterbuck had associated the sound with hunting, because the Rhodesian Hunters' Association had used vehicles, even though it had been four years since any vehicle had been used in hunting on Kazuma.

Another time Jim and I were crawling along the bank of a dry streambed, stalking waterbuck, when a leopard crossed about thirty yards ahead of us. We sat where we were for a long while, only continuing our hunt when we were confident the leopard was now far away. How wrong we were! When I caught sight of the leopard again—out of the side of my eye after just having crawled under a low bush—it was sitting quietly, staring directly at me. Jim and I slowly faced it, my rifle ready. The leopard was about four paces off from us—close enough that we could see every whisker as well as the detail in its eyes. Yet the leopard remained remarkably calm, so we did the same. At one point, Jim whispered, "Can I shoot it?" and I shook my head no. After what seemed ages, the leopard tired of us and scampered off. Jim immediately groaned, "Dad, why wouldn't you let me shoot it?" I patiently explained that we were not hunting the leopard; to shoot such a magnificent creature simply because we had accidently got close to it would be neither fair nor sporting.

---

[6] A waterbuck is a large antelope with a distinctive white circle resembling a target on its rump. The males have large forward-curving horns. A waterbuck can, at will, release a foul-smelling oil over its hair that repels water and predators.

In 1968, I invited a friend, Hans Otto (who was also a sergeant serving under me in the army) to live about a quarter mile from us and operate from Kazuma. Hans was passionate about wildlife and was trying to make a living capturing breeding sable and other game for the game farms now developing in South Africa. He readily set about selling the live animals I agreed to let him take. His income helped support all we were doing on Kazuma and—because I could now report a different stream of income from the land—allowed me to stop shooting zebra.

Hans captured most animals using a hired helicopter, newly developed dart guns, and drugs. However, because it was more economical, he also chose to capture young buffalo by driving his stripped-down, open Land Rover alongside each and grabbing them. Having offered to help him finesse this technique, I perched on his Land Rover's left front wing (fender), grabbing each young buffalo by its tail to slow it down until men could jump off and tie it up. This worked well, at least with the younger, lighter ones. I eventually managed to get the tail of a slightly larger one, which instead of running straight ahead, veered to the side, pulling me off balance. I fell behind the Land Rover's (still moving) front wheel and grabbed the first thing I could—the hot exhaust pipe. Thankfully, Hans slammed on the brakes, bringing the Land Rover to a screeching halt just before I would have had to let go and get run over by the back wheel. I was left with a severely burned hand. Soon after we abandoned this capturing technique as too dangerous.

One afternoon I took Shirley, Jim, and Meg on a relaxing game drive and invited our cook, Tomai, to come along. After driving through miles of open mopane woodland, we came to a rain-filled pan and decided to break for tea. Tomai went to get water while Jim, Meg, and I set about making a fire. A moment later, Tomai came scurrying back in terror saying, "*Ikona inyati emanzini*" (There's a buffalo in the water!). Now there obviously wasn't one, as the land in front of us had barely enough brush to hide a rabbit. I told Tomai to calm down. To prove that he had just imagined a buffalo and it was in fact safe to collect water, I started walking toward the pan—not even taking my rifle—with Jim and Meg slightly ahead of me, followed by a very nervous cook. When we got to the pan, I again reassured Tomai, "There you are, no buffalo." Just as I finished my words, a loud slurping noise emerged from a patch of tall sedges[7] in the pool—the only cover in sight. To my horror, an enormous bull rose out of the water and lumbered up to us. We froze, terrified. Jim, who was closest to the bull, fidgeted and I hissed, "Stand still!" It felt like forever before, thankfully, the buffalo ambled away.

Our lives at Kazuma were often peppered with "close but not too close" encounters with animals. Our open-air home, having permanently open doors and windows, was a delight. On one occasion I was wakened early by Bullet's muzzle in my face. Another

[7] Sedges are grasslike plants with triangular stems that grow in water or wet soil.

night, during dinner, the lights suddenly went out, though the generator we'd recently acquired for use with guests was still running. When I stepped out to investigate, I found that a bull elephant had walked between Meg's hut and the dining hut and snapped the wires. Yet another night, we were enjoying eland steaks when an owl flew in, grabbed a steak off the plate of a startled guest, and flew out. That owl, I am convinced, was one Jim had brought back from school as a baby and raised until it eventually went wild.

CHAPTER 24

ESPRIT DE CORPS

Purchasing Dimbangombe and Planning for Attacks

1968–1972

Those fighting the fire had wiped out all tracks of anyone who may have intended to kill me . . .

John Tswala, manager of Kazuma, standing beside the entrance. The sign reads: *Kazuma Range and Kazuma Research Station; No Entry Without Permission; Please—No Fires or Hunting; Private Game Reserve.*

One day Hans returned to Kazuma from Victoria Falls full of excitement. An old Afrikaner farmer, Van Tonder, wanted to sell his 6,000-acre farm, called Waterford, near Victoria Falls and return to South Africa. Hans had visited Waterford and found it to be perfect for his operations because the railway line ran through it. There was just one problem: he didn't have the money to purchase it—might I want to do so?

I knew that it would become increasingly hard to keep Kazuma wild; inevitably in those days a farm or ranch ended up as a community. The men in the labor force had wives and children who needed to be housed; soon a teacher was needed, then a clinic, then a football field and a farm store so they could buy essentials cheaper, and so on. I dreaded such development on remote Kazuma but could not see a way around it. Waterford, on the other hand—with its buildings, water, electricity, road, and railway—could be a perfect place to base everyone, in addition to being ideal for Hans' capture pens. There I could house staff families and simply have the men transported in and out of Kazuma.

I decided I would try to buy Waterford. Van Tonder, anxious to leave (fearing the mounting civil war), was willing to accept $20,000 cash. I did not have even a small fraction of that; however, I was determined to come up with the sum.

Two Americans, Bill Rutherford and Glen Gronquist, had contacted me a few years earlier at the suggestion of a mutual friend and visited me at Kazuma. They had crewed on the same bomber during World War II and remained close friends, annually visiting national parks in Africa with their wives. So enchanted were they with the wildness of Kazuma and the freedom to roam and track animals on foot that they stopped their trips to national parks and instead came to stay with us every year. With Rhodesia by this time sanctioned worldwide, they always brought needed items, from used fuel wing tip tanks for my plane to spare parts and saddles. I felt they might lend me the money.

I called Bill in Illinois, and he sent me the money immediately, saying that it was not a loan, but a gift in appreciation for the good times they always had at Kazuma. Hence, I acquired Waterford (years later renamed Dimbangombe after the river that runs through it). Because Waterford was to house Hans' operation as well, I formed a company to hold the land, with Bill and Glen as major shareholders (as events turned out, it was incredible luck that I did not put the land in my name, and the government was thus unaware I owned it).

By this time almost all farm and ranch homesteads in Rhodesia were fortified against attack with features like electric fences, bright lights, and radio connection to neighbors and the police. For my family this was not realistic. Our home at Kazuma was remote. We had no power for security lighting and even if we had a radio, our nearest neighbors—as well as the police—were hours away. As Hans was also one of my sergeants in the army unit I commanded, we decided we'd have to deal with any attack as best we could ourselves, always making sure one of us was with our families.

In case of an attack at night, Shirley and I made a plan with Jim and Meg, which we then tied to a plan for Hans' children. In the teak forest[1] between Hans' home and ours stood an old, distorted tree that the children called the "monkey puzzle tree." At the first sign of trouble, they were under no circumstances to try to come to us in our bedroom hut. Rather, they were to roll out of their beds onto the earth floor, slither out of their huts on their bellies, keep quietly to the dark, watch for muzzle flashes to know where gunfire was coming from, and only stand when they reached the forest. From there they were to make their way to the monkey puzzle tree and hide. They were not to move until all was over and the police, army, or Shirley and I came for them. This we rehearsed until

[1] Rhodesian teak (*Baikiaea plurijuga*) were slow growing, very hard wood trees that grew in vast forests on sand soils in both Rhodesias and Botswana. They provided most of the wooden props for the gold mines of South Africa and the ties for the early railways.

they knew the drill and could easily find their way to the tree.

Toward the end of our Kazuma years, in the early 1970s, we did have an incident that might have been an attack, though I was fortunately away with my family. Gilbert had tragically died in a vehicle accident shortly before, and my new manager, John Tswala (who had left what was now the Department of National Parks and Wildlife Management to join me), was at my camp. Hans and his family were up at theirs.

Around midnight John came out of his hut to pee and heard a dull explosion coming from my bedroom hut, followed by a ball of flames. All staff knew that in case of fire, it was critical to retrieve my gun cabinet kept right by the open doorway, but by the time John raced the fifty yards or so to the hut, the flames had climbed from my bed, up the mosquito net, and into the thatch roof. The heat was so intense it drove him back. There was a lot of ammunition in a large metal box by my bed that began exploding and Hans, hearing the noise, hid his wife and children and ran down to assist. He and the workers did what they could and were successful at keeping the flames from reaching other huts.

The damage was limited primarily to my bedroom hut—and the destruction of my beloved 7mm and .470.[2] To this day I don't really know what happened. Those fighting the fire had wiped out all tracks of anyone who may have intended to kill me, and if that was the cause, the culprit—by the simplest of observations—should have known I wasn't there. As we had no electricity or lighting, there was no way for a spark to have started the fire. John concluded that the dull exploding sound he heard was a petrol bomb. Years later, I spoke to a man who had been in the ZIPRA[3] army penetrating Rhodesia in that sector; he assured me the "attack" was not one of them. They knew I was isolated, but Gilbert had supported their party since our sugar farming days and had always spoken well of us.

During these game ranching days, I came to appreciate more than ever the value of teamwork, having made significant accomplishments despite the incredible hardships. Yet one incident made me realize I still had much to learn. While in the hospital following surgery on my knee, I took to wheeling myself up to the private room of a doctor friend, Jim Legget, who was terminally ill. He'd requested whisky, which I dutifully hid in my pajamas and smuggled into his room. As we sipped our drinks one evening—his nurse turning a blind eye—Jim asked me what I valued most in life. After reflecting a moment I said, "*esprit de corps*."[4]

---

[2] I later collected the burnt pieces of these rifles that had been so much a part of my life and sent them to Holland & Holland in London. They declined to rebuild them, stating that in the case of the .470, they didn't know the formula of the steel used and couldn't reharden it. Eventually I found two gunsmiths—a father and son team working from their home garage in New Mexico—who referred me to metal specialists in California. The specialists rehardened the action of the .470 and the father and son rebuilt both rifles beautifully. I am still using them to this day.

[3] ZIPRA was the guerilla army, Zimbabwe People's Revolutionary Army, that participated in the Rhodesian Bush War against White minority rule of Rhodesia. More is said about this in later chapters.

[4] *Esprit de corps,* from French, means "group spirit," referring to harmony and mutual understanding within a work group.

Jim replied, "Do you mean like the Hitler Youth[5] had?"

In a flash I could see the shallowness of my thinking. I realized that while some things can only be accomplished through good teamwork, it is the *purpose* of the work and the *character* of the leaders that determines the value of those achievements.

Game ranching still had me believing firmly that the removal of cattle, sheep, and goats would stop the degradation of soil and plant communities. However, almost by accident, I was about to make a breakthrough regarding the real nature of beasts in general, including humans.

[5] The Hitler Youth was a Nazi-organized youth movement; it was made up of different sections for boys and girls—the boy's branch was simply called the Hitler Youth.

PART VI INDEPENDENT SCIENTIST

CHAPTER 25

UNDERSTOCKED AND OVERGRAZED?

Navigating Science Versus Belief

1966–1967

I was beginning to suspect that wildlife managed for commercial productivity was not the solution to the deterioration.

John Acocks (1911–1979), who caught my attention when he pronounced South Africa understocked and overgrazed.

My obsession with reversing the degradation of the land fueled my desire to prove the viability of game ranching, and hence my role as independent scientist held steadfast alongside my roles of rancher, soldier, father, and husband. The data we were collecting from the game ranching operations by the WUS was encouraging but also puzzling. Initially, game populations were decreasing because only a few female animals survived to breeding age, due mainly to environmental or habitat degradation that had led to a lack of forage and cover (to hide from predators). Eventually, vegetation increased and enabled higher survival among younger-aged animals, but the vegetation tended to be woody, older plants showing less overbrowsing; very few young plants were successful in establishing themselves. Bare soil was increasing, not decreasing, which in turn led to decreased rainfall effectiveness and with it, environmental degradation. I was beginning to suspect that wildlife managed for commercial productivity was not the solution to the deterioration.

In my independent quest for solutions, I read every wildlife management, ecology, and soil science textbook I could get my hands on. I also read hundreds of research papers on veld management, many published by the Society for Range Management, an international organization I had recently joined. Though most of the research stemmed from countries where range science was most advanced (primarily North America and South Africa), none of the research addressed the ecological puzzle I was struggling with—and no research papers provided information that I, or any rancher, could use in day-to-day management.

Range specialists, both academic and practical, tended to determine the condition of the land by the species of plants present. Those they considered good, useful, and/or "native" reflected good land condition. Those plants that were considered toxic ("invasive"), useless, and/or "non-native" reflected poor land condition. On the other hand, as an ecologist, I assessed the same land first by the condition of the soil surface between plants. What percentage of soil between plants was bare, hard capped, eroding, shedding rainfall, or allowing water to evaporate out of the soil? Second, I assessed the land by the diversity of life on and within it and, third, by the age structure of the plants, which indicated whether their populations were expanding or decreasing. To me, these multiple factors were a closer reflection of the overall health, stability, and productivity of the land than "playing God" in judging what species were good or bad.

I found the ecological terms *native* and *non-native* bureaucratic rather than scientific. For billions of years, species of plants and animals have been spreading around the world but establishing only where the habitat suited them. Biological communities have even developed on newly formed volcanic islands thousands of miles from land. It makes no sense to set a date and call everything arriving after that date "non-native," and in many cases, "invasive." Once any species arrives, it either finds suitable habitat and establishes itself, or it does not and dies out. The new species can only "invade" if the habitat is suitable for it. That's not to say we should deliberately introduce new species, especially predators. Predators can cause mayhem for years until the entire biological community changes, having lost and gained species.

I reached out repeatedly to my range science colleagues, raising concerns over both the extent of the degradation and the rate at which it was increasing. Yet their responses were always the same. When not blaming rainfall, they blamed the public—mainly commercial ranchers and communal farmers subsisting largely on cattle and goats—for not following their advice. As scientists, they claimed, "We know what to do but we can't get the public to do it because it always involves reducing livestock numbers."

But were they right?

One morning while visiting Ian Henderson on his ranch, Doddieburn (where game ranching started), I skimmed through a copy of *Farmer's Weekly* on his coffee table.

The magazine, published in South Africa, had an article titled, "Understocked and Overgrazed"[1] covering the work of botanist John Acocks. I immediately found Acocks' thinking to be intriguing and sensical. I phoned him and arranged to visit him in the South African Cape, more than 1,000 miles away.

I decided to drive to the Cape, but I was so hard up at the time I couldn't afford more than my fuel. I slept in my car each night, eating food I had brought with me. I'd never travelled that far south in the winter, and the nights were unexpectedly cold and uncomfortable. Having neglected to pack even a jersey (sweater), I had to feign hardiness in my shorts and short-sleeved shirts.

Acocks was a delightful botanist who had studied the steady decline of perennial grassland and expanding desert in the Karoo region of the Cape.[2] All that was left of the Karoo grasslands were the towns developed around the former freshwater springs (*fonteins*), each heralding the Afrikaans name of an animal that once grazed there—*Elandsfontein, Springbokfontein, Oliphantsfontein,* and so on. These towns were now surrounded by parched Karoo desert and dry, woody plants; almost all the eland, springbok (an impala-sized antelope), and elephants were long gone.

In our long discussion, Acocks reasoned that the game animals of yesterday had grazed differently than the livestock of today. The literally uncountable game animals, he said, had severely grazed all the vegetation equally, giving no preference to one plant over another. In comparison, the sheep—in much lower numbers—selectively grazed, feeding on their favorite plants until they killed off those plants, then turning to other plants until eventually all that was left were the least palatable bushes and annual grasses (perennial grasses having largely disappeared). Based on these assumptions, Acocks proposed a "non-selective grazing" system in which each farm was divided into sixteen fenced "paddocks" and livestock forced to eat all the plants in one paddock before moving to the next. This, he concluded, would ensure an equal grazing of all plants, eliminating unfair competition among the grazed and ungrazed plants for water and nutrients.

Though intrigued, I questioned some of Acocks' reasoning. Was the problem really the unfair competition among species of grass plants? I thought about this concept when applied to species of animals. I—and many others—had often observed a lack of any game over large areas of land, only to come upon one species and find several other species feeding nearby—which we suspected was for mutual predator avoidance. For example, after seeing nothing for miles we would suddenly see zebra and almost immediately kudu, puku, impala, waterbuck, and eland. However, with crocodiles there appeared to

---

[1] From "Non-selective grazing. Part 1. Understocked and overgrazed!" by D. Howell (*Farmer's Weekly,* December 28, 1966: 31).

[2] The Karoo is a semiarid geographic region that covers the Eastern Cape, Western Cape, and Northern Cape provinces of South Africa. The Karoo is best defined by its vegetation, which consists of assorted succulents and low scrub bushes spaced from one foot to several feet apart.

be a noticeable predator benefit. While on Northern Rhodesia's Kafue River with Len Vaughan one day, we arrived at a spot where it appeared the rocks were moving but they were actually crocodiles waddling into the water. We fished there, catching 235 fish in a short while—we literally could not pull a hook through the water without a fish grabbing it. Somehow the mass of crocodiles and the abundant fish populations were connected. These and other experiences caused me to change my views about competition in nature. Species both above and below ground do not live in competition with one another, but rather as a biological community of thousands or millions of species in great synergy. If we only look at a certain species through the lens of "survival of the fittest," we see competition. On the other hand, if we look at everything through a community lens, we see the opposite of competition: cooperation, reciprocity, and synergy.

As an independent scientist, I was increasingly departing from the opinions of other ecologists who referred to many different ecosystems, implying that each one operated differently. Just as I could not envision plant ecology separate from animal ecology, I could only envision one global ecosystem—or biosphere—containing a great many environments, all operating under the same ecological principles. (When books such as *Fire Ecology*[3] later appeared, they made no more sense to me than if one were to publish a book titled *Chainsaw Ecology,* for both fire and chainsaws are management tools—there is no "ecology" of a tool.)

The independent scientist in me had another concern based on my observations. The healthiest grasslands I had seen in Northern Rhodesia (now Zambia)—from the Kafue Plains, Luangwa Valley, and Bangweulu Swamps to the Mweru wa Ntipa Reserve—hosted thousands of buffalo and lechwe as well as thousands of elephants. However, these herds never remained in one area long enough to consume absolutely everything. Many severely grazed grass plants tended to be present close to water sources, where a dense sward of runner grasses would develop as bunch grasses decreased. However, elsewhere the grass plants were tall, dense, and abundant, and few of them overgrazed.

These herds moved every few days, accompanied by many pack-hunting predators. Several times in my Northern Rhodesian Game Department days, while walking amongst the game, I counted more than forty lions in a day. Consequently, the buffalo ran together in herds numbering 3,000 to 4,000 head or more—I simply could not count them; there were buffalo as far as I could see. On the fringes of the Bangweulu Swamps, my godfather Eustace had reported seeing a single herd of lechwe he estimated to be continuous for more than twenty miles.

The predators, I felt, played an important role, although I did not yet fully understand that role. However, in 1966 American ecologist Robert Paine more fully articulated the

[3] Today there are many publications with "fire ecology" in the title—the book being referenced here is *Fire Ecology* by Henry A. Wright and Arthur W. Bailey (New York: John Wiley & Sons, 1982).

importance of predators in his landmark ecological paper, "Food Web Complexity and Species Diversity."[4] Paine's experiment showed how removal of one numerically minor "keystone species"—in this case a starfish that preyed largely on mollusks—could cause the collapse of a large and diverse community of many species. Later, biologists in East Africa would argue that wildebeest, not predators, were the keystone species in what they coined the "Serengeti rules."[5] These biologists failed to fully grasp Paine's insight; key predators ruled there, too, for more reasons than simply being an apex predator like the starfish, as I was still to discover.

---

[4] Robert T. Paine. "Food Web Complexity and Species Diversity," *The American Naturalist*, Vol. 100, No. 910 (Jan.–Feb., 1966), pp. 65–75, The University of Chicago Press.

[5] The Serengeti Rules: The Quest to Discover How Life Works and Why It Matters, Sean B. Carroll (Princeton, NJ, Princeton University Press, 2016).

CHAPTER 26

GOLDEN HOOVES OF THE SHEEP

Seeing Livestock as the Solution, Not the Problem

1967

Over all my years of condemning livestock . . . I had not been open to the possibility of livestock being beneficial to the land.

(*left*) The grassland pictured here appears lush when looking across it but looking straight down (*right*) reveals what I found on the Howells' farm: grass plants too widely spaced, exposing bare ground between them that is sealed against the infiltration of water, free passage of air, and the sprouting of seeds. Where their sheep trampled and broke the seal, plants grew closer together.

As an independent scientist I was now learning more than ever before about the relationship among soils, plants, and animals. Although some of that learning made me question Acocks' theory that selective grazing was the cause of the Karoo desert expansion, Acocks assured me that farmers using his sixteen-paddock, non-selective ("forced grazing") system were seeing positive results. I was there to learn and asked if I could visit ranchers who were following his advice. Acocks introduced me—over the phone—to nearby farmers Len and Denise Howell, who immediately invited me over to stay with them and see their land. So began a long friendship.

After my first night in a warm bed, Len and Denise took me on a tour of their farm. The morning was bitter cold, made a little less so when they insisted on lending me a jacket. Everywhere we drove, the Howells pointed enthusiastically to the increased grass growth and return of species they hadn't seen in years. I had no reason to doubt their observations that—judging by species—the grassland was improving. However, as noted previously, I now assessed improvement not by species alone, but by overall ecosystem function, starting with the condition of the soil's surface. If the surface was bare and sealed, soil organisms couldn't breathe, and water couldn't soak in. In a healthy grassland, the cycling of water and mineral nutrients is unimpeded; plant and animal communities above and below ground are complex and healthily populated; and the conversion of sunlight energy in green, growing plants is maximized. By these measures, the reappearance of lost grasses on the Howells' land stood out as only one positive sign in a still deteriorating landscape. The grass plants were still too widely spaced, exposing too much bare ground between them. Much of that bare ground was capped or sealed against the infiltration of water, free passage of air, and the sprouting of seeds.

At one point when Len stopped for me to open a gate, I noticed an area about fifty yards away that looked remarkably different. I immediately left the gate to investigate, digging my fingers into the loose, dark soil. Len and Denise came over to ask what I was so excited about. I pointed out that the ground there had closely spaced grasses, and the soil between the plants was covered with a protective layer of litter (mulch formed of dead and trampled leaves). Diversity was noticeably higher, too, and the ratio of leaf to stem much better, making the grass appear lusher. All this even now, at the height of winter! It was just what I had been seeking; it was the closest I'd seen to the healthy grasslands present under near-natural large game herds threatened by pack-hunting predators. Something had happened here that was significantly different than what was happening on the rest of the ranch. What was it?

The Howells confirmed that their sheep often crowded together in this corner of the paddock as it provided the most protection from wind and snow. I had an epiphany! The sheep's *behavior* could do what was needed to reverse the deterioration that had obsessed me from my earliest days as a game ranger. The combination of fencing and foul weather had done what I had seen predators do to large herds—cause the animals to bunch up periodically. This unpanicked, crowding-together-behavior also, I realized, helped ensure constant movement, as once the area was saturated by the sheep's dung and urine, they sought to leave it (no animals like to feed close to their own feces).

We were failing to achieve this outcome in our game ranching operations, but why? There were several reasons. We had fewer numbers of game animals than had once existed; and fences, roads, and other developments restricted the animals' movements and their ability to gather in large numbers. We also had severely reduced the populations of pack-hunting predators.[1] This revelation was a turning point for me. From that moment on I knew that *livestock,* so long vilified, were not the cause of the land deterioration, but the source of a solution if we could just figure out how to replicate routinely what I had seen on that one corner of the Howells' farm.

Over all my years of condemning livestock—based on my childhood observations while surveying dusty, dying lands occupied by cattle, goats, and donkeys with my father, and later what I learned at university—I had not been open to the possibility of livestock being beneficial to the land. I had been wrong! But why had someone not realized this thousands of years ago? Why had livestock—whether it be cattle, sheep, goats, camels, horses, or donkeys—always been blamed for the loss of biodiversity and environmental degradation, no matter how pastoralists, and lately scientists, aspired to do better? As I would later discover, many pastoral peoples had known there was a link between their animals and the health of the land. Scottish shepherds, for instance, had referred to the "golden hooves of the sheep," without which the land suffered, though they couldn't explain why. Early Dutch farmers in South Africa had a saying in the same vein, *"Hammer the veld [with hooves] to sweeten it."* Yet, scientists hadn't investigated these observations.

The corner of the paddock that boasted such a healthy community of grasses certainly did not care what beasts had impacted it; instead of sheep, it very well could have been cattle or even wild game. The key was not so much the "what" as the knowing *why* and *how* this type of impact happened.

On my long drive home, my mind was spinning. I had never worked with livestock apart from ploughing with oxen as a boy and participating in the frenzy of cattle dip day on my uncle Guy's ranch (when animals would be sorted via the cracking of whips and forced to plunge into and swim through a chemical bath to remove the ticks responsible for a variety of diseases). As for fencing, I had nightmarish memories of the many thousands of wild animals killed by the fences put up to control foot-and-mouth-disease or the spread of the tsetse fly.

Like my fellow ecologists and range scientists, I had believed we had to protect land from livestock, most importantly from high numbers and large herds. The need to minimize numbers of livestock had been the subject of thousands of research papers, textbooks, and the teaching of leading universities. Contrary to Acocks' thinking, land degradation was *not* due to unfair competition (too few animals overgrazing plants selectively while

[1] Had the observations of nineteenth century English trader and adventurer George Mossop been published at that time it may have hastened my understanding of the critical role the hooves of excited animals—and their dung and urine—played in soil health. As scientists, our focus had always been on animals' mouths—their grazing and browsing. In *Running the Gauntlet: Some Recollections of Adventure, Second Edition* (Johannesburg: G.C. Burton, 1990), Mossop writes of wildlife herds in the 1870s that took hours to pass, their hooves raising clouds of dust that could be seen for miles. The land then was covered in grass and marshy "vleis" full of reeds and open water. Yet fifty years later, when the large herds had disappeared, the land had dried up.

permitting other plants to outcompete the favored ones). However, his conclusion that the number of animals—too few (not too many)—is connected to land degradation was correct. Now the independent scientist in me was charged with a newly discovered desire to learn how we might (re)introduce the golden hooves and reverse the desertification.

PART VII ECOLOGICAL CONSULTANT

CHAPTER 27

RATIONAL, NOT ROTATIONAL!

Grazing Periods, Planning Processes, and Simplicity

1966–1968

Voisin had studied some 200 years of grazing in Europe—which had generally followed a systematic rotation through several pastures—and concluded that rotational grazing consistently led to the loss of biological diversity in plants and soils.

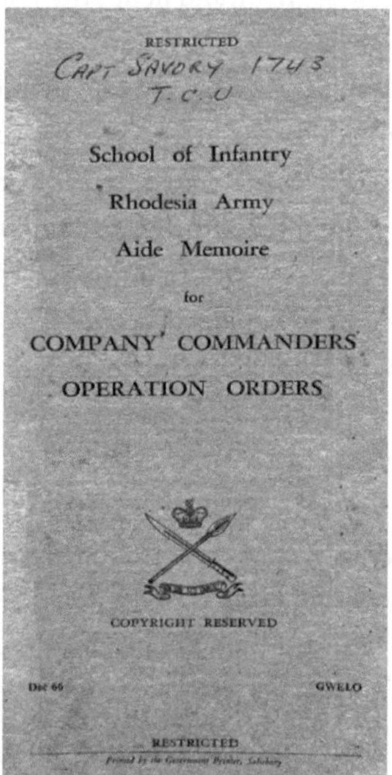

Titled *School of Infantry, Rhodesia Army, Aide Memoire for Company Commanders, Operation Orders,* this aide memoire inspired me to create one for grazing planning. *(Image provided from my personal archive.)*

For years, I had publicly opposed commercial cattle ranchers because of the perceived damage their animals caused throughout Rhodesia. This was reinforced by the popular opinion that people were to blame entirely for the degradation happening on Rhodesia's communal lands—they were running far too many cattle, goats, and donkeys. I too had learned to think like this at university and often said that I was prepared to "shoot all the bloody ranchers, if necessary, to save my country." Then one afternoon, after my return to Bulawayo from the South African Cape, there was a knock on my door. Before me stood a well-known Matabeleland ranching couple, Evelyn and Edward Rushmore. Change was about to happen.

Surprised, I invited the couple in for tea, asking why they had come to see me—their enemy. Edward stammered badly so Evelyn did all the explaining. "Allan, we've been following the conflicts brewing with the government, the research stations, and you. We deeply care for our ranch and have religiously attended every field day at Matopos Research Station to learn how to do things right. We've tried to implement all the recommendations of the researchers and the government extension advisers, but our land is just getting worse, not better."

I listened intently, identifying with their frustrations. Evelyn concluded by saying that what they had learned of my views made sense to them, and hence the visit. Then she asked, "Allan, will you help us?"

For the first time I felt I was conversing with cattle ranchers who genuinely loved their land as much as their animals. For Edward and Evelyn, ranching was not all about money, as I had believed it was for most cattle ranchers. We talked for a long time. It was clear that we shared a deep concern over the deterioration of their land, but I told them I had no answers. I pointed out that my fellow scientists were either denying the problem or blaming ranchers and communal farmers for ignoring their advice (which consistently involved reducing livestock numbers). I also pointed out that for thousands of years as far as I could see, no one had solved this problem of land deteriorating where humans ran domesticated animals, particularly in the world's vast, seasonally humid regions. In the end, I agreed to provide my services as an ecological consultant, provided they understood that I did not know what to do but would work with them until we found answers. Our collaboration would be a case of the blind leading the blind. Together we would figure out how to best manage cattle in a way that would maintain and hopefully even improve the health of their land.

The Rushmores became the first of my "client-collaborators" in what ultimately became an international consulting business. Within days, Colin Bickle, who ranched near the Rushmores and was married to Jenny, the daughter of George Rudland (the minister of agriculture whom I had clashed with in attempts to meet with Prime Minister Ian Smith during my Kapombe Farm days), came to ask if I would take them on as a client under the same terms. I agreed.

The Rushmores, Bickles, and I had to start somewhere, so I began with John Acocks' non-selective grazing system. On each ranch we sectioned off sixteen paddocks, combined the cattle into one herd, and began to rotate that ranch's one big herd through the paddocks as Acocks had prescribed. I visited frequently to monitor progress, riding out from Bulawayo on my motorbike to keep my costs down. We saw almost immediate change on both ranches, as had Acocks on the Howells' land. When prolonged overgrazing stopped, plants grew. And most new plants grew where the physical impact of the cattle was highest—by the gates, corners, and anthills, which attract cattle. However, I was not seeing the same improvement on the ranches as a whole—too much soil remained bare between the plants. In addition, I was grappling with how to merge livestock needs with wildlife needs as both the Rushmores and Bickles loved the wildlife on their ranches. There was also the problem of coordinating livestock movements with the demands of the crops grown on the two ranches. We had much to learn, yet very little to go on.

Years before, I had bought Andre Voisin's book, *Grass Productivity*. Voisin was a French pasture scientist whose work involved small properties, animals confined by fencing, and a perennially humid climate. At the time my mind was mostly on elephants and large African game; I saw no connection and had put the book aside. Now, in working with livestock, I dusted it off to see what Voisin had to say. He had studied some 200 years of grazing in Europe—which had generally followed a systemic rotation through several pastures—and concluded that rotational grazing consistently led to the *loss* of biological diversity in plants and soils. So, what could we learn from him?

All the textbooks and scientific papers I had read referred to "land" being properly grazed when animal numbers were low but overgrazed when there were too many animals. Voisin's work corrected in my mind a major error in range science thinking—that *land* could be overgrazed. Only plants could be grazed or overgrazed, not land—the latter is a complex biological community entailing soil, rocks, trees, other plants, algae, millions of insects, and small organisms. Correcting this seemingly innocuous bit of syntax forces us to distinguish between what happens to some plants and what happens to soil, litter, microorganisms, and the entire biological community.

In addition, none of the literature defined *overgrazing*. Voisin didn't either, but he did describe how overgrazing occurred. If grasses were grazed repeatedly for too many days, they would be overgrazed. Or, if animals returned to re-graze grass before the grass had sufficiently recovered and regrown both leaves and roots, the grass would be overgrazed (and in this case even more severely damaged).

Now it was beginning to make sense to me how pastoralists—moving their herds of livestock over the land—contributed to land degradation *gradually* over centuries while modern ranchers, guided by range scientists, were destroying their land much more rapidly. Pastoralists kept their animals concentrated as they herded them each day; in turn fewer plants were overgrazed. In contrast, modern ranchers were releasing their herds into spaces controlled by fences—which restricted livestock's movement—then leaving their animals there (scattered rather than concentrated) for long periods, believing this prevented overgrazing. In both cases, the land was still degrading, but the time it took to do so under more concentrated versus more scattered animals was significantly different.

I didn't unravel this enigma until more than a decade later when I observed Pakistani pastoralists herding their animals in a manner that apparently had not changed for thousands of years. I realized that over centuries their main purpose in herding was to protect their animals from pack-hunting predators. But because the animals were protected from predators, there was little *herd effect*—the agitated trampling that breaks up sealed soil surfaces and helps create litter, as I had seen at the Howells' farm. It turns out that this behavior change was just as important in a herding situation as the timing was to the establishment and regrowth of plants, and thus to the health of the land.

The grazing system regarded as the best in Rhodesia at the time was the "three-herd, four-paddock" system developed by my longtime friend, Oliver West, along with another well-known scholar of the veld, C.E. Tidmarsh, and others. This rotational grazing system was prescriptive in that it never varied from year to year, no matter how much the weather, season, and thus growing conditions varied. It was favored by researchers throughout southern Africa but, as I had observed (and the Rushmores and Bickles too, on their ranches), it still led to a loss of biological diversity and gradual desertification.

Voisin emphasized that the *timing* of plants being grazed was what governed whether plants were overgrazed, rather than the number of animals grazing the plants. Thus, to avoid overgrazing, Voisin would make sure that animals didn't return to a paddock until the grasses had recovered and regrown from the previous grazing. This insight proved to be much more revolutionary than it sounds. It was also more practical than Acocks' approach. Although Acocks' system resulted in longer recovery times than previous practice for those plants actually grazed, the recovery periods were far longer than needed for most grass plants. With recovery periods too long, and/or too few animals to graze the area, many of the perennial grass plants remained ungrazed, their leaves and stems turning grey as they slowly broke down in direct sunlight through chemical oxidation, which led to their premature death. This is something that did not occur in more reliably humid Europe, where Voisin had done his research.

Voisin had developed a very simple way for farmers to plan how to graze their animals. He based this on the prevailing rate of grass growth, which varied according to season and weather. Voisin called his planning process "rational grazing"—meaning thought out, or planned, each season—rather than "rotational grazing," which implies sticking to a routine, regardless of the variables. Voisin's results were well documented in increased milk yields (a more sensitive and immediate indicator of livestock productivity than weight gain) and increased land productivity and stability.

Guided by Voisin's research, on both the Rushmores' and Bickles' ranches I chose to stop rotating the cattle based on the estimated time it took the herd to graze out each paddock. Instead, we planned herd moves based on the time it might take for *any* severely grazed plant to recover, factoring in the differing daily growth rate of plants and the number of paddocks we were moving the cattle through. Voisin's thinking helped me realize that any formula for *rotational* grazing must eventually fail—no matter how adaptable or flexible—because nature's elements don't subscribe to a routine schedule. Voisin also helped make it clear for us that some form of *planning process* had to replace rotational grazing *systems*.

I continued carefully observing the results of our changes, often walking across the land during rain to see what happened to the soil surface as the rain fell. If I saw a night storm brewing in the direction of the two ranches, I'd jump on my motorbike and be on the land by dawn to watch rainwater flowing off the land, noting how the runoff related to the soil surface and that in turn to the physical impact of the animals.

Voisin's planning process—involving some calculations and a map, calendar, and pencil—was a major step in the right direction. However, I concluded that his process—while adequate for relatively simple pasture situations in Europe with its more stable climate and better distribution of humidity—was not as suitable for our more complicated situation in Africa. Rainfall was more erratic, and wildlife and crops all had to be synchronized with one another. What we needed was a simple planning process that could handle greater complexity.

Faced with a seemingly impossible jumble of clues, I needed a planning process that encompassed answers to a multitude of questions:

- How could we manage livestock not just to produce food, wool, or hides, but as a tool with which to reverse habitat destruction for all life (including humans)?
- How could we control the movement, feeding, and breeding of livestock?
- How could we increase livestock numbers safely without damaging land?
- How could we change their behavior daily to have a beneficial physical impact (herd effect) on vegetation and soil?
- How could we minimize overgrazing of all plants, ensuring sufficient recovery time for even the most vulnerable in a complex plant community in which all species play a role?
- How could we prevent many plants from remaining ungrazed and thus oxidizing in sunlight and dying prematurely with too few animals?
- How could we do all these things while also catering to wildlife needs and crops on the same land?

All these questions had to be answered in the context of our erratic rainfall and dramatic seasonal variations, not to mention the process had to be simple enough for the average person (often with minimal education and literacy) to grasp. A tall order.

I had no assistance from universities or sympathetic agencies; it was just me—an ecological consultant riding on his 90cc motorbike to save fuel cost—and a couple of ranch families. I resorted to taking long walks in the bush near Bulawayo. I would pause under a tree to puff my pipe, wander, then pause again, all the time mulling over possible solutions. Cycling endlessly in my head were the few clear facts. Every way people had ever tried to manage livestock—even the most ancient livestock-based cultures—had not catered for the complexity involved. That had led to some level of environmental degradation and, in the seasonally humid/arid areas, to desertification. Yet somehow livestock, if only we knew how, were key to reversing this.

Not wishing to reinvent the wheel, I investigated other professions that might have learned how to plan successful outcomes in complicated, ever-changing situations. The range science profession had shied away from the issue entirely in the belief that reducing animal numbers and developing ever better grazing systems would solve everything. Whole libraries on management, even in business schools, provided no books on planning techniques that were simple, practical, and relevant to our situation. Business executives, for example, had developed management systems to handle accounting, inventory, and so on—but these systems were based on the premise that the general context would remain relatively stable and predictable. They had not attempted to create a management system to run their business as a whole (which was neither stable nor predictable). Among all the professions I looked into, only military institutions appeared to have solved this problem in a serious and methodical way. Battles throughout history had generally involved complicated, unpredictable, and fast-changing situations. The victorious side typically was led by commanders who excelled at planning, but how?

Starting with my school days at Plumtree onward, I had gone through extensive infantry officer training. The curriculum—from Britain's highly regarded Royal Military Academy Sandhurst—boiled centuries of military experience in battlefield situations down to a pocket card for planning, referred to as an "aide memoire" (French for "memory aid"). The aide memoire broke down any situation requiring an officer to plan (whether it be an attack, a withdrawal, or a river crossing) into simple bullet-point steps. No matter how tired, stressed, or confused an officer may be, he only needed to follow the aide step-by-step to devise the best possible plan in real time. We officers were all trained in the use of the aide memoire and always carried one with us.

The human mind tends to seize up when tasked with multiple options simultaneously; this paralysis occurs even more so under stress (as farmers frequently experience). Try, for instance, to count every motorbike, car, and truck coming through a busy intersection at the same time. It's extremely difficult for most people to do. In essence, the aide memoire's method of planning acknowledges this; no human mind can consider all the variables simultaneously—so don't try. Instead, the aide memoire helps one deal with each segment, one at a time.

I realized I had surfaced what I was looking for to help with grazing plans—I needed an aide memoire that could handle any farm, ranch, or pastoral situation in any environment and season, good or bad. To do this, I first identified the steps a rancher, farmer, or pastoralist needs to take in creating their grazing plan. I then put the steps in order so that they built on one another. For example, some of the simple steps I identified for the Rushmores and Bickles were:

1. note herd information (such as numbers);
2. note when the herd can't be in certain paddocks (because of planted crops, ground-nesting birds, etc.);
3. note the paddocks requiring special attention (such as the need to be grazed heavily by the end of the growing season to prevent fires, or before frost reduced feed value);
4. determine the length of recovery needed by the first severely bitten plant so that all plants recovered; and
5. determine the length of the grazing periods so no plants were overgrazed when animals were in any one paddock.

In using my aide memoire for grazing planning, I could now focus entirely on one step at a time before moving on to the next—and not feel overwhelmed by so many factors all at once.

Yet I faced another hurdle in adapting the concept of an aide memoire to grazing plans. Battles are generally fought for a short period; however, anyone running livestock needed to plan for months, and in some cases more than a year. In addition to longer-term planning, I had to consider different sizes and qualities of paddocks, as well as changing growth rates of plants. I decided I needed a graphic representation of my plan—a visual to accompany the aide memoire. Hence, the birth of a grazing planning chart. Using this visual, I could record information from each step and then see the whole pattern of events and issues clearly before planning the actual moves of the animals.

The top of my grazing planning chart was divided into months, with smaller divisions accounting for time down to the day. The rows down the left side represented paddocks or sub-divisions of land. At the base of the chart and on the left and right sides was space for various planning figures and calculations, including the plant recovery periods and the grazing periods derived from them. In the central body of the chart, covering both days and paddocks, I noted and color-coded problems and needs—an orange line across Paddock 3 in April for instance, might indicate poisonous plant danger, a brown line across Paddock 4 in October might indicate antelope fawning, and so on. Then, within the context of all these factors, I planned livestock moves. The Rushmores and Bickles immediately implemented my planning process, using the aide

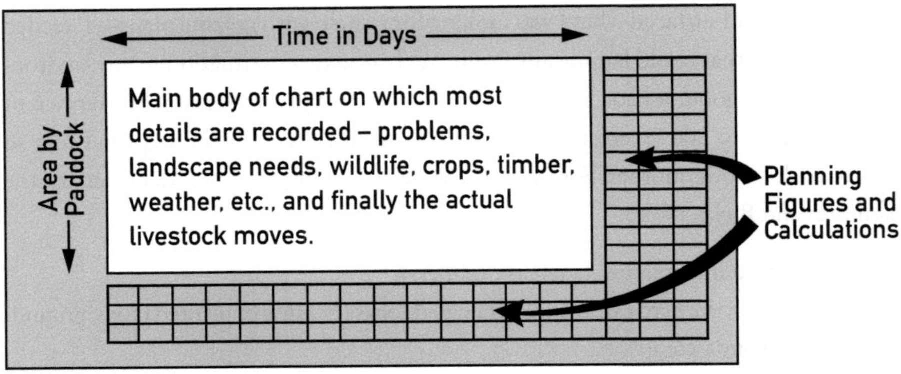

The layout in principle of the grazing planning chart.

memoire and grazing planning charts I designed for them, and it wasn't long before we could all see improvement on both ranches. Not only did we feel a sense of joy in this, but we also felt less stress—we could now plan many months in advance with ease. We started to increase animal numbers to the benefit of the entire biological community, witnessing both animal and land health improve. It was exciting to realize we were healing land using livestock which, after 10,000 years of contributing to desertification and the demise of civilizations, had become a pivotal tool for creating healthy habitat for wildlife *and* humans.

At this point, I had to stand back and reflect on how my newfound revelations affected my romance with game ranching. Was game ranching still the best way to save game and reverse land degradation (though I could now see that game ranching alone was continuing to result in habitat degradation)? Wouldn't it be beneficial to use livestock as a tool as well, if managed properly through my grazing planning process? Might we reap the benefits of both wildlife and livestock, without conflict?

There would be many (and still are many) who argue that using livestock to restore environmental health should not undermine the ideal of "re-wilding"—enabling natural processes to repair and restore degraded land. And what about the jungles—the so-called "lungs of the world"—that ranchers cut down to raise beef? And how about the tsetse fly, brucellosis, and predator control?

These were all good questions I had yet to address and would not do so until many years later; a vital piece of the puzzle was still missing. However, my initial euphoria over solving a 10,000-year-old puzzle sustained my optimism. It was perhaps good fortune that at the time I had no idea of how fraught with institutional resistance and difficulty implementing my solution would be. After all, in 1903 a pair of bicycle mechanics by the last name of Wright finally got an airplane off the ground; within twenty-five years after that, Lindburgh crossed the Atlantic; and a mere thirty-five years after, Gagarin orbited the earth!

CHAPTER 28

SCIENTIFIC CHARLATAN! SNAKE OIL SALESMAN!

Gaining Popularity—and Mitigating Doubt

1968–1970

Right away the board agreed that all farmers should have the opportunity to hear what I had to say and judge for themselves.

Mike Fairlamb made it possible to share my ideas with farmers throughout the country.

Believing then, as I still do, that our discovery of a way to reverse desertification over vast regions of our habitable planet ranked up there with the achievements of the Wright brothers, I drove to Salisbury to see Oliver West, now chief of the Department of Research and Specialist Services in Rhodesia's government. Oliver was our most respected and knowledgeable scientist in veld and pasture management, and as I've described earlier, a family friend. We had a long discussion in which I explained what I was doing and what I had learned. I shared how the Rushmores, when using his three-herd, four-paddock grazing system, unfortunately only saw increased deterioration. This couldn't have been easy for him to accept, even though a report had already gone public on the soil erosion occurring in the research plots on which the grazing system had been based.[1] Nonetheless, I told him

[1] The report that highlighted soil erosion under Oliver West's three-herd, four-paddock grazing system was the *Phillips Report to the Rhodesian Government*, published in 1962.

it was my hope that the government research stations, extension services, and he would take over the development of my work, helping its success spread both in the country and internationally. He agreed that this could be done, however, he would first have to submit my grazing planning process to experimental testing. This would take several years, assuming they could even get the funding to do it.

This was a stalemate I had not expected. The planning process, I said, was just that—a *process* to plan complicated, ever-changing, unpredictable situations using all the knowledge available. It didn't make any sense to try to prove experimentally whether a planning process worked; by definition only planning could succeed in situations of great unpredictability as every businessman knew. What we could do as scientists was to document the results achieved when using the planning process. I was documenting the results myself through a rigorous monitoring procedure and I suggested government scientists could do the same if that was required for official acceptance. Oliver agreed but insisted this argument would not convince others within the government—which meant he wouldn't get the budget for it. It was a sad moment for me to part in disagreement with a man I respected and loved and to whom I owed so much for my very interest in the veld. We parted friends, and over the next few years, when my character and motives were attacked repeatedly by my fellow scientists, Oliver West, always a gentleman, never joined in the ridicule.

Failing to get government to advance our breakthrough had me relying on my own meager resources once again. I did not know how I was going to do it, but my parting words to Oliver were that in the two or so years he'd said it would take him to get the funds and the approval to do his experiments, I would get the whole country moving.

Soon after my meeting with Oliver I met Mike Fairlamb. He had been commissioned as an officer in the same territorial army battalion[2] and worked for the Bulawayo Farmers' Co-op. Mike and I became immediate friends, sharing many of the same ideals and a warped sense of humor, amongst other shortcomings. Mike knew all the Matabeleland farmers and of course the Rushmore and Bickle families, so he had heard of our efforts using cattle to heal the land. Over many beers in the officers' mess, I explained how I had adapted our military planning, which Mike understood immediately. He mentioned that he was on good terms with Colin Bickle's father, Albany, who was chairman of the co-op. We also both knew Major Cedric French, who worked for the co-op and was one of our company commanders. Mike felt that, with Albany's and Cedric's support, he might persuade the board to finance a series of talks to farmers around the country in the spirit of broadcasting new developments. Via this channel, he wouldn't have to negotiate official recognition with the co-op.

---

[2] I detail my army days in Part VIII, Irregular Warfare.

In due course I got a call from Albany Bickle. Would I address his board? Yes! Right away the board agreed that all farmers should have the opportunity to hear what I had to say and judge for themselves (there was already talk among government researchers that my work was not supported by research and should be ignored). And so began my speaking tour around Rhodesia, sponsored by the Bulawayo Farmers Co-op.

Farmers and ranchers found what I had to say exciting, rarely hearing good news after years of harangues about overstocking the veld. Requests for my help skyrocketed, which I tried to accommodate as best I could. With such demand it made sense for me to make a serious business of my consulting, and hence I began charging realistic fees. I also found myself driving more than 4,000 miles a month, often at night, between clients. My consulting practice not only expanded in Rhodesia but into Swaziland (now Eswatini), South West Africa (now Namibia), South Africa, and Bechuanaland (which in 1966 became Botswana). This—and an increase in financial stability—led me to buy a plane (a secondhand, single-engine Piper Comanche 180) and learn to fly. Subsequently, many farmers put in simple, bush-cleared air strips where I could land.

Not only did the purchase of a plane help significantly in traveling great distances for my consulting business but it also made toggling between three homes more manageable. One house I rented in Bulawayo, near where Jim and Meg attended boarding school (I started flying them back to school from Kazuma at the end of school holidays, marching them straight to the plane from a bucket washdown in our camp sink). I rented another in Salisbury, where I could stay cheaply when attending Parliament.[3] The third was my family's main home, our Kazuma bush camp. When not staying in one of these three homes, I was likely in the WUS hunting camps.

My plane also enabled me to get much-needed fuel for my vehicles at Kazuma. Petrol was strictly rationed due to the sanctions, but aviation fuel was not. With the help of American pilot friends, I acquired two secondhand, wing tip fuel tanks from a scrapped Comanche 250. Although the tanks weren't certified for my plane's lower-powered engine, and war being on, we forged papers and installed them. The plane flew better with these tanks and could now carry ninety gallons of fuel, giving me greater range and safety. I compared this to my ration coupon, which allowed me to only buy one drum of diesel fuel every month (and even less petrol for my Land Rover). Subsequently, I started running my Land Rover on aviation fuel.

I would simply fly twenty minutes to Victoria Falls' airfield, top off my plane with ninety gallons of fuel, fly home, and syphon most of the fuel out. Days later I would fly into the same airfield and again fuel up. Although I had to log all my flights, the airfield's system was not sophisticated enough to flag a Comanche for burning so many gallons of fuel!

---

[3] I detail my Parliament years in Part IX, Parliament.

Remote as we were at Kazuma, we could not hide from Rhodesia's increasingly worrisome political situation. We were right on the border of Botswana, which had become an independent country in 1966 and harbored no sympathy for the Rhodesia led by Prime Minister Ian Smith. I once needed to discuss a matter with the Rhodesian police at the border post at Panda ma Tenga and said I would fly over as it was close to my Kazuma home and airstrip. Via phone (at this point we also had a party line phone) I asked the officer in charge, "Where is your landing strip? My map doesn't show one."

The officer responded, "It's right by the police station—you can't miss it!"

From my plane, I saw the long obvious airstrip by the police station, however, it appeared to be on the wrong side of the border, so I circled. Perhaps, I pondered, I was looking at a veterinary fence for foot-and-mouth disease that I did not know about. There was no other strip to be seen—and the policeman had been very clear—so I landed.

I realized my mistake too late when men, armed with AK rifles, surrounded my plane. Thankfully I was dressed in rugby shorts and a faded army jacket with no insignia or rank on it. I got out of the plane, mustering a smile and friendly greeting. Feigning calm, I asked how their crops and cattle were, all the time implying that I lived nearby in Botswana. Then I asked, "Is that the Rhodesian police station just over the fence?" They confirmed so. I eyed the wind and runway. As long as no one blocked me before I could get into my plane, I could make it with a downwind takeoff. I bid goodbye (again, ever so friendly), got back into my plane, gunned the engine, and quickly took off. Little did my newfound armed friends realize they had just let a reasonably large fish go—a Rhodesian member of Parliament and commanding officer of the Tracker Combat Unit (both roles I go into further detail on in subsequent chapters).

Meanwhile, the government began to disseminate research discrediting my work. A friend, Mick Gammon, who had been at school and university with me as well as often on the same cricket team, was now a researcher at Matopos Research Station. He published a peer-reviewed paper stating that his "experiment" failed to support my claims. According to the report, he had laid out a series of one-acre fenced paddocks and rotated a herd of two steers through them to impact the land. The land hadn't improved at all, and hence he declared my process faulty.

I traveled to Matopos to have a look at Mick's setup. As Mick and I neared his experimental plots, I asked, "Where is your herd?" Mick pointed to two steers standing in the shade of a thorn tree on one of the one-acre fenced plots. Astonished, I explained that two animals are not a herd and cannot behave like a herd, even when standing together for shade. Eventually I think Mick understood that his experiment was a far cry from that of a herd of 200 animals moving through several 100-acre paddocks. The same stock density (two animals per acre) applied in each case but produced an entirely different effect on the land. Unfortunately, his paper had already been published—and that paper is still being cited today, sixty years later, as evidence that I am a "scientific charlatan."

My competition as a consultant consisted of "reputable" scientists thriving on government and university salaries and offering free services. These scientists considered me a snake oil salesman trying to make money. I had to rethink my business. I needed to outsmart free, reputable experts with no economic-vested interest (but a greater and more insidious vested interest of professional egos). I wanted ranchers and farmers to commit to at least three years of collaboration in which I could continue to consult plus observe, learn, and measure the results.

I developed a two-stage process. In the first stage, when ranchers and farmers approached me for help, I generally said yes. My fees for this first visit were low, and depending on the size of their operation would cover my time for two to four days with them. During those days I assessed their situation and how they could better use livestock to increase land productivity. I planned the layout of possible fencing, water sources, and other infrastructure (most properties, in addition to livestock, had tobacco and maize, though I also consulted on properties that produced tea, coffee, cotton, and/or managed wildlife). I taught each client my planning process, providing them with grazing charts and an aide memoire.

My personal aim during this first visit was to decide whether I wanted to continue working with the client. At the end of my visit, if they were not people I wanted to work with (fortunately a rare occurrence), I left them with a good start and wished them well. If I decided I wanted to continue to work with them over a longer period, I urged them to keep going and contact me if they needed further help at any time. *Or* they could consider retaining my services.

Inevitably, clients would ask what my retainer cost would be per year, and I would quote my fee, which, depending on the farm or ranch size, amounted to a few hundred to a few thousand pounds per year. My fee included installing monitoring sites (transects)—from which I could gather information on any changes and take fixed-point photos—and at least two visits a year.[4] I knew my fee would shock people who weren't used to paying when government advice was readily available and free. So, as I quoted my fee, I allowed no time for a reaction, but immediately said, "That shocked you, didn't it!"

The response was always the same, "Yes, that's a lot of money!"

To which I responded, "Yes, it is! So, I have a suggestion: How about we go double or quits? Meaning, during a three-year-minimum commitment, I will double your stocking rate (or carrying capacity) with documented improvement of the land. If this doesn't happen, I will return every penny. But if it does happen, you will pay me double the fee."

[4] Any number of visits beyond the first two that were deemed necessary to ensure success I did at no further cost. If I had to visit many times, I reasoned it was because I had failed to advise clearly and hence always incurred the cost.

Compared to the prospect of doubling the production of any ranch, my "high" fee was trivial. Every rancher or farmer to whom I made this offer committed to the full three years; never did a client renege on our verbal agreement nor need to ask for their money back.

Doubling the stocking rate on ranches was easy to do because stocking rates, up to this point, had been based on the traditional belief that overgrazing was caused by too many animals. Stocking rates were decided by reasoning like, "take half the grass by year's end and leave the other half as insurance to prevent overgrazing" and were thus always very low. Now that we could reduce or eliminate the overgrazing of plants, there was no need to take only half and leave half (which in any case did nothing to stop the overgrazing). My clients, with grass now growing ahead of and behind their moving cattle, produced much more forage and comfortably fed double the number of animals.

At the end of each three-year commitment, all my clients renewed our agreement. Although we routinely doubled the stocking rate, I never did double my fees. Tragically, the more ranchers that engaged my consulting services in Rhodesia and beyond, the more I seemed to offend academics, veld management experts, and government extension services. The official reaction to the very idea of using livestock to reverse desertification or of increasing livestock numbers was bitter and extreme. Many people (even those who would today be deemed environmentalists) regarded my shift from "game ranching consultant promoting wildlife utilization" to "ecological consultant promoting livestock as a tool for land regeneration" as treason. I was, after all, helping the "enemy"—cattle ranchers. Even my longtime friend and mentor Ray Dasmann, the Fulbright Scholar who had been so crucial to our early work in game ranching and now occupied a position at the International Union for the Conservation of Nature (IUCN) in Rome, refused to see or talk to me again when I told him livestock could help solve the problem of habitat degradation.

Unfortunately, ridicule and doubt exist to this day (and I anticipate will continue to the end of my life), with climate change "experts" and anti-meat activists determined to label all forms of livestock grazing as irredeemably evil. I had unwittingly thrown "a pork chop into their synagogue" as Mike Fairlamb gently joked.

CHAPTER 29

"PERFECTION OF MISMANAGEMENT"

Testifying for the Farmer and Against the Government

1967–1971

"Oh my God," Bob exclaimed. "How have so many of us gone so wrong for so long?"

Bob Vaughan-Evans standing in dense perennial grass (1969); two years earlier, this was largely bare soil in which Bob stood, lecturing ranchers that they needed to kill the trees because the trees were preventing grass from growing.

In a further attempt to meet the escalating demand for my consulting services, I began gathering and consulting whole groups of ranchers at a time. The only stipulation was that all ranchers within a group had to agree to openly share details of their operations with the others. The group, with no exceptions, had to work as a team, helping each member as needed. Their meetings with me would move from ranch to ranch, ensuring every ranch within the team received attention. Working in this way I was able to help ten or more ranch families at once, saving both time and costs.

One day a Namibian rancher, Malcolm Kroon, called to ask if I would represent him in a court case. The South African government, which had governed Namibia (former German South West Africa) since World War I, was expropriating his Tsumis Ranch to form a "bastergebeid" or an area where "Coloreds" (mixed race people) could live and run their livestock. The government had offered to compensate him for his land at a price he believed was far below market value, and he was suing. He was finding it difficult to get professional representation against government and a senior advocate in Namibia had suggested he call me.

I knew nothing about Namibia, and my calendar was booked three months in advance, so I declined. Malcolm persisted, suggesting I cancel some of my less critical obligations, and I again refused. My clients took priority. Desperate, he then asked, "Allan, what would you charge if you did represent me?" To continue to dissuade him, I quoted a ridiculously high fee. To my surprise, he immediately offered to double that fee. I finally agreed to visit over an upcoming Easter weekend because he clearly was desperate for help. Once I met Malcolm in person and learned how unfair and bullying the South African government was being, I agreed to represent him.

In due course we all attended court in Namibia's capital city, Windhoek. The South African government had set their price for Tsumis Ranch based on other properties they had purchased on a "willing buyer, willing seller" basis. Malcolm argued that his ranch was of higher value due to the quality of the soil and the state of the vegetation. To counter this, the government introduced evidence from a ranch they had recently purchased, which they claimed was in even better condition. In fact, their professional range scientists had described the ranch as "perfection of management." I said I could not comment without seeing this so-called superior ranch, so court was adjourned for a day to allow me a visit.

I spent a long day taking measurements and photos all over the ranch and had the photos developed and printed overnight. When court resumed the next day, I advised Malcolm's lawyer to recall the state's expert witnesses. This he did. The scientists responded angrily to our questions, which were childishly simple. *Did they know what soil erosion was? What types of soil erosion were there? Was there any soil erosion of any type on the ranch?* The witnesses were certainly knowledgeable on the various forms of soil erosion yet were adamant that there was no erosion of any kind—wind, water, sheet, or gulley erosion—on the ranch being called into question. It was, they reiterated, "perfection of management."

Then it was my turn in the witness box. I was asked the same questions, to which I testified with the data and pictures I had collected the day before. I pointed out that the ranch *appeared* to look good—a sea of waving grass—but looks were deceiving. Beneath this grass was almost entirely bare, eroding soil dominated by only one perennial—an arid environment grass species that relied on bare soil to establish itself. I backed this up with pictures of grass roots sticking into the air half an inch or more due to soil loss. I cited American research stating that a dime thickness (about one millimeter) of soil loss was equivalent to over twenty tons of soil loss per acre. And I wrapped up my testimony rather tactlessly, concluding that this so-called exemplary ranch was a "perfection of *mismanagement*."

Malcolm won his case, and the government was forced to pay him a fairer price, but ripples from what became known as the Tsumis Case lasted for years. The first was that Pete Hanssen, the Namibian government economist who had appeared in court arguing against me, invited me to have coffee with him. He subsequently left government service and joined me in helping to train Namibian ranchers (later I helped Pete and his family move to the United States as the situation in Namibia worsened). Malcolm Kroon engaged me to consult on another ranch he had in the encroaching desert, and later also on several properties in the South African Cape. On the other hand, the South African government, certainly in part because of this case, banned me from setting foot on any university campus in South Africa for more than twenty years.

In yet another repercussion, I had been engaged by the government of Lebowa, one of South Africa's Bantustan territories[1] for the Sotho people in the northern Transvaal. There, we had successfully increased cattle numbers, and the improvement in land health was clearly seen for miles—contrasting with that of South Africa's bare ground and encroaching thorn trees. I was due for one of my visits to Lebowa in 1978 when their officials informed me my contract had been terminated. They had no choice, they said. The South African government had insisted they cut all ties with me. If they chose not to, all of South Africa's aid to Lebowa would cease.

During these awkward years of intense and—to me—inexplicable rejection and obstruction, I nevertheless gained a valuable ally in my oldest friend from school days, Bob Vaughan-Evans. I had given a talk to a crowd of farmers in the Que Que (today Kwekwe) area of Rhodesia, where Bob served as a senior soil conservation officer. Bob attended my talk and afterward walked me to my car. He was clearly upset. When I asked what was wrong, out it all came. He told me it had taken him eight years of working with these farmers to convince them to accept conservative stocking rates, as well as the clearing of invasive woody vegetation. The investments were many—from contour ridging (swales) to gulley control—to deal with the degrading land. He now felt that I had discredited him in public, making all he had done meaningless. Why would I do that?

Bob said he was told I was bitter against government, and that I only cared about money. None of this matched the friend he knew from our school and university days, but he didn't know who or what to believe. I was shocked. I suggested that maybe we could spend a day together to quietly discuss it. The following weekend, Bob drove down to Bulawayo. Rather than rehash our former conversation, we simply walked out on the land near my home on the edge of town. There were no cattle on this land, which appeared to be healthy savanna grassland. I then encouraged Bob to take a closer look.

[1] The Bantustan territories were ten former territories designated by the White-dominated government of South Africa as pseudo-national homelands for the country's Black African (classified by the government as Bantu) population during the mid- to late-twentieth century.

Between the plants, I showed him the bare, eroding, hard-capped soil. I also pointed out the wide plant spacing and lack of age diversity in the plants—they were nearly all old with few if any new seedlings. Because there were no livestock (just a small number of hares and a few solitary duiker), only a few plants had been grazed. None of the dead stems and leaves had been trampled to the ground to provide litter (mulch) covering the soil. I showed Bob how chemical oxidation was replacing biological decay of the dead leaves and stems, which were slowly turning grey rather than breaking down rapidly. In conclusion, though it appeared healthy when looked at superficially, it was a gradually dying grassland.

Bob and I both earned our degrees from the University of Natal. His degree was in agriculture; mine was in zoology and botany. Bob had gone on to do years of training under veld management professors. I had never attended a veld or range management lecture in my life. The criteria through which his constituents and he judged healthy land and management were entirely different from the criteria I used. For hours we discussed the differences—what he had been taught at university versus what I had been learning on the ground; how water cycles functioned or failed; what effective rainfall meant; and finally, the role of fire and of animals in such landscapes. I will always remember the aha moment when, as we were standing under a marula tree, Bob realized the truth. "Oh my God," he exclaimed. "How have so many of us gone so wrong for so long?"

From then on Bob remained a staunch ally and dear friend, assisting me behind the scenes in invaluable ways. He frequently phoned to warn of pending government actions against me. He readily sent me copies of confidential documents informing me of government experts' views and moves to discredit my work. Sadly, decades later, he was murdered in post-Rhodesia violence. Although Bob had been actively supporting farmers who were being forced from their land by the government, Bob's murder was not associated with his work but with a robbery gone awry. Upon hearing a noise in his home, he jumped out of bed to the surprise of a burglar, who axed his head, killing him instantly.

CHAPTER 30

MORE "TRIALS" AND TRIBULATIONS

An Attempt to End the Controversy

1967–1975

Swift assured me, "with all the know-how of the United States" that the land we were looking at was technically beyond reclamation.

The Advanced Project on Liebig's Ranch: thirty-two paddocks radiating from a central point, as seen from my plane (1970). Within a few years this land became the most productive on the entire ranch, producing five times as much meat per acre.

By the late 1960s the number of ranchers experiencing impressive results was reflected in the enthusiasm for my talks. If I talked to ranchers and farmers anywhere in Rhodesia, the hall was generally overflowing. When Minister of Agriculture George Rudland or his officials spoke, the room, in stark contrast, was largely empty. This did not help my already strained relations with the Ministry or its extension officers.

One evening in 1968, while staying with a rancher client near Salisbury, I tuned in to the evening radio news to find that Rudland was publicly challenging me to a national trial. Shortly afterward a reporter from the *Salisbury Herald* called, asking for my response. I gave it, but he said he could not possibly print it. I then said that he would just have to wait; if the minister wanted to make constructive suggestions, he knew where to find me (we both had offices in the same parliamentary building).

Several days later, Rudland (who first clashed with me in my sugar farming days) asked if I would come and meet with him, which I did. We first agreed that from there on we would deal with each other directly rather than lob grenades through the media. This paved the way for a more friendly conversation. Rudland acknowledged that his daughter and son-in-law (my original clients, Colin and Jenny Bickle) had reported impressive results using my "grazing ideas."[1] However, his advisors were rejecting my claim that I could double the cattle stocking rate and increase profit while improving the land. Instead, they insisted that land under the management I advocated would deteriorate and profit decrease. Rudland proposed that government run an official trial to determine who was right.

To end the controversy "once and for all," as Rudland expressed in his proposal, we agreed to compare the results of my planning process with the government's prescribed grazing system. This was precisely what I had suggested to Oliver West (Chief of the Department of Research and Specialist Services) as the only way we could "test" my ideas—by comparing results side by side. Each would have a block of land as a demonstration site because, as I had informed Oliver West, we could never submit a planning process to experimental replication.

To up the stakes, I offered to double the stocking rate on not just one but two blocks of land. In addition, on the second block, I would show how ranchers could pay for costly infrastructure (fencing, water points, and so on), without having to borrow capital (the money would be generated from the land as the land grew increasingly productive). Hence, my two "trial" blocks became known as "Rich Man Savory" and "Poor Man Savory." For the former, capital was provided to construct twenty-one paddocks. For the latter, capital was provided to construct five paddocks with the potential to increase as the land improved and profits accrued.

Rudland, on behalf of government, promised that government policy, training at the agricultural colleges, and extension service advisers would endorse my work provided the Savory blocks validated my claims of greater profit while improving rather than damaging the land under a doubling of the stocking rate, compared with the government-recommended-practice "control" block. On the other hand, if my claims were wrong, the government would require me to close my consulting practice and stop misleading the country. I felt this to be fair. We sealed the deal with a handshake, which for me was as

[1] I had no name for my grazing planning process at this point as we were still developing it. People took to calling it "Savory Grazing" but I preferred to not have my name used, so I called it "short duration grazing." However, in 1978 range scientists in Texas plagiarized my work, labeling it the "short duration grazing *system*" (later called the Wagon Wheel System) and converted it to just that—totally ignoring the essential planning process. When I learned of this, I changed the name to "Savory Grazing Method" to differentiate it and then, finally, to "Holistic Planned Grazing," which is what it has been called to this day.

binding as a legal agreement, but oh, how naïve I proved to be. A legal document sealed with the Rhodesian Arms might not have served me any better, given the spirit of the times and what I was learning about the behavior of "expert authorities" and institutions when confronted with what they viewed as heretical ideas.

This challenge was named the Charter Estate Grazing Trial (or the Charter Trial) as Rudland already had approval from the London-based board of Charter Estates[2]—which operated a large cattle ranch close to Salisbury—to conduct the trial on their land. Their general manager, David Worthington, was a highly respected cattlemen who also provided the cattle and management. Government assigned Dr. Stan Parsons, an agricultural economist from the local university, to document the profit of each block. Dr. John Clatworthy, the head of Grassland Research in the Department of Research and Specialist Services (and another friend of mine from school and university days), was assigned to monitor the land.

We immediately got to work setting up our blocks. Using maps, soil and vegetation samples, and aerial photographs to guide us, we initially laid out three "equal or comparable" blocks on largely open grassland. Within a couple of months, however, the grazing charts on the two Savory blocks (recording forage offtake of every move the cattle made) showed a sizeable difference in their productivity. This meant that despite our concerted effort to identify equal blocks, we had still misjudged them and had to realign their boundaries. The government "control" block thus became two blocks (Charter 1 and Charter 2), each with four paddocks. The Poor Man Savory block remained as it was, and the Rich Man Savory block was reduced in size, leaving it with fifteen paddocks. Both Savory blocks continued to run double the cattle numbers of the government blocks.

Within a year of launching the Charter Trial, Mike Gawler, Ranch Manager at Liebig's Ranch, asked if I would advise the ranch on the management of their grazing. Liebig's Ranch—spanning 1.25 million acres—ran more than 60,000 cattle. Having Liebig's Ranch as a client was an opportunity to test the grazing planning process in ways not yet done, and I immediately said yes. I knew that their land was deteriorating as I had camped, hunted, and done game surveys there with WUS (which leased the hunting on the place). I also knew the ranch was—by law—having to plunge-dip 12,000 head of cattle daily to comply with veterinary regulations for tick control. This required the animals to walk nine miles on average to and from dips. Not only were they losing weight and condition, but they were also creating eroding trails on the land.

I sought to test a new fencing layout, proposing that the fence radiate out from a dip and water point at the center, which would eliminate the cattle's long walk to water

[2] The Charter Estates was purchased in 1946 by a group of British investors. In 1948, the investors recruited David Worthington to manage the sprawling estate. At its peak, it sustained 11,500 head of cattle, 4,000 or more of which were breeding cows.

and dip as well as large investment in multiple watering points. I had also designed a new type of fence I wanted to try that wildlife would be less likely to get caught in and die. If we tested this cheaper, wildlife-friendly fencing layout and design on the worst area of Liebig's Ranch and it worked (as well as my grazing planning process) we should then be confident to proceed over the entire ranch.

Another favorable aspect was the relationship I enjoyed with Mike Gawler. Mike's open-mindedness was one of the reasons I had befriended him years before. His son, John, and my son, Jim, were the closest of friends at Plumtree school, which had brought us together on numerous occasions. It was because of his trust in me that Mike agreed to speak with Graham Ballard, Managing Director of the Liebig Company. And it was Graham's trust in Mike, I believe, that quickly got us the approval from Liebig's London headquarters to move forward using the new, untested ideas I was proposing.

I knew that if I selected anything but the worst land on the ranch, my critics could say I had started with good land. The southern parts of Liebig's Ranch were therefore an ideal place to start. In the pioneering days this region had been prime productive grassland supporting large numbers of game, including roan and sable antelope. Despite roan and sable being declared royal game (not to be hunted for any reason) both had completely died out due to the deterioration of their habitat. So bad was the deterioration in the south that the original ranch headquarters had been moved further north to Towla. In addition, rainfall was erratic and extremely unreliable, averaging twelve inches but sometimes as low as four inches.

I had personal experience with how bad the southern parts were—it was the area where I had struggled in my game department days to keep two horses alive on 20,000 acres while doing a game survey (there was little to forage but shrubs, tree leaves, and twigs). I also remembered (again while I was with the Game Department) the time a team of international scientists headed by American Dr. Lloyd Swift visited Rhodesia and East Africa to report on the extent of desertification. When I showed the team the land I now wanted to use, Swift asked, "Do you see any hope of reclaiming this land?" I replied affirmatively, saying I believed we could, if we could get our government to address the problem and stop blaming droughts. Swift strongly disagreed, assuring me and all present that, "This land is technically beyond reclamation with all the know-how of the United States."

Because I also wanted my efforts on Liebig's Ranch to be in the nation's eye, and not just a local ranching effort, I invited the chairman of the Rhodesian Cattle Producers' Association, Glyn Richards, to visit the site. I offered Glyn five pounds (then about ten dollars) for every perennial grass plant he could find on a 100-mile drive. There was literally no perennial grass. Thus, on Liebig's Ranch we selected 4,000 acres of what was some of the worst, most degraded land in Rhodesia.

While testing the new fencing layout and design, I also intended to increase the cattle numbers to see just how many we could run before detecting signs of land deterioration. I knew that if I called this effort a "trial" and succeeded in causing its failure (I was explicitly aiming to use ridiculously high animal numbers), my opponents would publicly proclaim the whole approach a failure. So, I called it the "Advanced Project" instead.

One of the first things we did in launching this Advanced Project in 1970 was install new fencing, constructing thirty-two paddocks that radiated out from the center like a wagon wheel (as shown in the photo that opens this chapter). This many paddocks gave us more control over the timing of cattle moves than the ranch had ever had before, which in turn enabled the project area to grow an enormous amount of forage. We doubled the livestock numbers immediately, and before the year was out, we tripled them. Within three years, the land in our Advanced Project became the most productive on the whole ranch, and by its eighth year it was producing five times as much beef per acre as the surrounding 200,000-acre control area. There were higher calving rates and weight gains, and the total capital cost in infrastructure (fencing, water, and dip tanks) was less than a pound (roughly one dollar and eighty cents) per acre! In addition, hunters soon learned the Advanced Project was attracting game and would go there to find the freshest game tracks to follow. Due to our wildlife-friendly fencing, herds of zebra, kudu, and wildebeest were easily moving and concentrating where the large cattle herd had been a few days before. Considering this success, we began planning the grazing and realigning the fencing for the rest of the ranch's 1.25 million acres.

Two years later, Bob Rutherford, the owner of the nearby Igava Ranch (and son-in-law of Charter Estates Manager David Worthington), asked to visit the Advanced Project. Keen to work with us after touring the project site on Liebig's, he agreed to host a second Advanced Project on Igava. Rutherford's ranch averaged forty inches of rainfall—more than triple the rainfall of Liebig's. This second Advanced Project would give me the opportunity to show what was possible in a higher—but still very seasonal—rainfall environment.

The results on Igava Ranch were, in fact, like those we achieved on Liebig's Ranch: higher profit from greater productivity of cattle and land. However, with higher rainfall—and thus more mineral leaching, which results in what we commonly called "sour veld"—Igava also experienced lower individual animal performance. This was a problem that would take us a few more years to solve (though not urgent because Igava Ranch was still more profitable than before).

Alongside the two Advanced Projects at Liebig's Ranch and Igava Ranch, the Charter Trial ran for six years until 1975, when the government decided to end it with the point—we all agreed—proven. Closure was to be announced at a public meeting in Salisbury, chaired by Dr. Piet Booysen, Professor of Veld Management at the University of Natal. I knew Dr. Booysen (he had attended the university at the same time as I), and I looked forward to his visit.

I thought it would be useful for Dr. Booysen, while in Rhodesia, to visit our Advanced Project on Liebig's Ranch, so the day before our closing meeting, I flew him to the project site. While walking across the land with Liebig's managers, Dr. Booysen repeatedly uttered, "My God, this is amazing—I never would have believed it without seeing it." On our flight back, I eagerly anticipated the next day's meeting, knowing that in addition to our impressive success on Liebig's Ranch, every person involved agreed we had proven both the Rich Man Savory block and Poor Man Savory block more profitable than the government grazing system.

The public meeting ended in a shambles. In his opening speech, Dr. Booysen, to my astonishment, declared that there was no evidence that planned grazing was any better than the grazing system recommended by government. I—along with others—stared at him in disbelief, wondering what had happened overnight. Dr. Parsons' report and data showed greater profitability on both Savory blocks compared to the two government blocks. Dr. Clatworthy's report, showed only a minimal change in plant species on the Savory blocks, when in fact dramatic changes were expected. He noted:

> *The Charter Trial was intended primarily as an economic comparison of different methods of managing cattle on veld grazing . . . although no major differences in grass cover showed up in our measurements, there were great differences in the general appearance of the various systems. Under the conservative stocking rate of the Charter system [the government block], the grass grew tall and dead top hamper accumulated so that periodic burning was necessary to remove this. On the rotationally grazed plots*[3] *[the Savory blocks] the heavier stocking rate kept the grass short and there was no build up of dead material. This gave the sward a very healthy appearance.*

On both Savory blocks, we had Dr. Parsons reporting greater profit, and Dr. Clatworthy acknowledging the improved land health, stating that his analysis by species alone had not shown the improvement all could see. And toward the end of the meeting, when manager David Worthington was asked by a member of the public what he now intended to do on Charter Estates, he replied that he *absolutely* intended to continue with my planned grazing process (which he did, not to mention his son-in-law Bob Rutherford continued using the process as well on Igava Ranch). Yet the Rhodesian government, specifically the Ministry of Agriculture and George Rudland, deemed the Charter Trial

[3] Dr. Clatworthy was never part of the planning and wasn't aware of the distinction we made between rotational and planned grazing.

a failure (many academics still cite it as a failure).[4] In the minister's eyes, despite the stated aim of the trial, I had failed to prove that my planned grazing was any better than the government's recommended grazing system. The ministry soon redoubled its efforts to shut my ecological consulting business—and me—down.

Shortly after, Dr. Parsons resigned from his university position, left for South Africa, and joined me as a partner, helping to service my South African clients. Dr. Parsons convinced me I should start spreading my work through teaching, believing training courses would reach more ranchers than consulting alone ever could. Hence, we launched our first courses in grazing and financial management.

[4] There was one piece of evidence in which the government blocks were higher, and that was the individual performance of the cattle, that is, each animal in the government blocks put on more weight and the conception rate was higher than in animals in the Savory blocks. However, individual animal performance was not something the Charter Estate Grazing Trial was testing (profitability was being tested). Fewer fatter animals was leading to less profit and less benefit to the land. These were problems I was yet to solve entirely although the Liebig's Advanced Project also showed higher individual animal performance as well as profit.

CHAPTER 31

THE "GRAZING GAME"

Shifting Mindsets . . . and Animals

1967–1978

What if our plan was the reverse of how ranchers normally react to droughts?

Despite my detractors, there was still considerable public interest in my work, as shown in this *Illustrated Life Rhodesia* cover story (December 15, 1971).[1] (Cover text reads: *Rhodesia 2001 A.D.: a barren desert incapable of supporting life—this is Allan Savory's grim vision of the future. But, he says, it is a vision which need not become reality.*)

I was becoming thoroughly disillusioned with academia. I had been taught at university to expect harsh criticism of my work as a scientist, but never to take it personally as criticism of me or my motive as a person. However, now that I was the one questioning the work of established scientists, the scientists very much took the criticism personally, responding by drumming me out in any way they could.

[1] *Illustrated Life Rhodesia* was published between 1968–1978. It was banned and ceased publication in 1978.

At a well-attended ecological symposium in Pretoria in 1967, I presented a paper titled, "Efforts in Rhodesia to Apply an Ecological Philosophy and Practice to the Human Environment to Avert Semi-desert Formation." I argued that our management was leading to habitat destruction and that habitat was vital not only to wildlife but also to humans as a species no less dependent on its environment. I asserted in my conclusion that poor land leads to poverty, social breakdown, and even war. To my astonishment, my presentation received a standing ovation. However, of all the papers presented, mine was the only one not published.

Government efforts to shut down my consulting and research intensified, both in Rhodesia and South Africa. In South Africa, following the Tsumis Case, I was informed at the last minute that a planned meeting with farmers had been cancelled because the government had told them that if I even spoke in a private home, all farmers in the district would lose their government subsidies.

Rhodesia had only two agricultural colleges at this time: Gwebi College of Agriculture, near Salisbury, exclusively for Whites, and Mlezu College of Agriculture, near Que Que, exclusively for Blacks. Gwebi students were traditionally allowed to invite lecturers of interest, but George Rudland, still minister of agriculture, banned the students from extending an invitation to me. While the principal at Gwebi, fearing retribution, honored the ban and distanced himself, the principal of Mlezu—an open-minded Afrikaner named Fritz Meyer—did not (in addition to serving as Principal of Mlezu, Fritz was also a respected rancher and a close colleague of my friend Bob Vaughan-Evans). When Fritz asked me to take on Mlezu College as a client, I pointed out that government would never approve of him paying my consulting fee. Fritz shrugged it off, telling me to just leave that to him.

Fritz had no problem defying the government, and thus we had a few great years working together. He was an innovator; everything he did impressed me. He managed crops, fish, and dairy cattle, integrating them all to spur amazing productivity despite the poor granite soils of his area. From the air, Mlezu stood out as an oasis.

Not long after Fritz had engaged me, I was contacted confidentially by another unexpected client—an economic branch of the government itself! The civil war had been steadily worsening, and people were beginning to feel the sting of worldwide sanctions. Most Rhodesian farmers were astoundingly creative and resourceful managers, surviving whatever was thrown at them. Many were not born farmers but had left various professions to become tobacco farmers because it was so lucrative, supporting a lifestyle that included polo, yachts, and private planes. When the sanctions came, many of these farmers successfully diversified into cattle, sheep, coffee, cut flowers, and other enterprises. However, some farmers began to fail. Abandoned farms were vulnerable to becoming guerrilla bases—an outcome the government certainly didn't want. It was thus critical that farmers stayed on the land. To do this, a new quasi government-funded "bank"—

the AFC (Agricultural Finance Corporation)—was formed. The AFC took over when other banks would no longer assist bankrupt farmers. In turn, AFC-supported farmers were tightly controlled but had the funds they needed to support their families and keep operating. Much to my surprise, the AFC started sending their clients to me, paying my fees in the hope that I could help these farmers stay in business.[2]

From my regular clients, my AFC clients, and a few hard-up friends I helped pro bono, I learned much about the most important element of a farm—the mind of the farmer. If a farmer was mentally buoyant, accepting responsibility, never blaming, and never complaining, you couldn't hold them down. But if mentally defeated, lazy, or tending to blame others, no amount of financial aid or advice seemed fruitful. Interestingly, regular clients were generally of the former mindset, whereas my AFC clients and the friends I helped for free were often of the latter—less malleable—mindset. Most of those ultimately failed despite having the AFC's help.

Around 1975 Gumela, a small ranch in the beautiful Matopos Hills (where, as I've previously mentioned, Cecil Rhodes chose to be buried) came up for sale and I bought it. Gumela's simple home was built on a gigantic granite rock amongst boulders. It was a quiet and beautiful place where I could envision living out my final years. Tom Orford, a friend from my game department days, had resigned from the department and his prospects appeared shaky. I invited him to have a look at Gumela and its lovely home in the rocks. If he liked it, I said he could live there free and ranch for himself (for the moment I needed to stick to my consulting business and research, not to mention that I was increasingly tied up politically and militarily).

Tom favored the idea, but he needed to buy cattle and had no money. Borrowing money was by then almost impossible in our embattled economy, and I felt lending money to a friend was the best way to destroy the friendship. So, Tom and I formed a partnership on the cattle. I agreed to raise the money, but the partnership had to repay it with interest. The bank would not lend the money to the partnership on the security of the deeded farm—nor accept cattle as security—but agreed to provide an overdraft to the partnership, if I deposited a similar amount in a CD for their security. I made the deposit, and Tom, with his wife and two children, began ranching on Gumela.

For better or worse, in the mid 1970s a genuine drought coinciding with an insect outbreak tested all I had learned. In the Kezi District (in which Gumela was located), the normally low rainfall was particularly bleak, creating drought conditions made much worse by the mostly noneffective rainfall. This coincided with a heavy outbreak of a caterpillar we called "armyworm."[3] Millions of them ate every leaf they could reach,

[2] My first AFC-funded clients were Charles and Rose Loxton, owners of a small ranch close to the Matopos Hills south of Bulawayo. To my horror, my success led to their deaths. Had I not kept them and their children on the farm, they would not have been there when a guerrilla gang attacked their isolated home and murdered them.
[3] Armyworm is the larval stage of the *Spodoptera* exempta moth.

whether grass, bush, or tree. Ranchers in the area began panicking, selling cattle as quickly as they could. Two of my clients there, Afrikaner brothers Ben and Boet van Vuuren, owned four ranches in the district and were doing well with their grazing planning. However, they were worried, as cattle on ranches all around them were dying and hordes of armyworms swarmed over their land, too. They called for advice.

The van Vuuren ranches were being affected by circumstances outside of Ben and Boet's control—and this required that we entirely replan. With nowhere to send cattle, we needed to ensure the survival of as many cattle as possible while also sustaining the land improvements made to date. The only plan I could come up with was a risky one—were they willing to consider it? The brothers, their backs to the wall, said yes. I then posed the question: *What if our plan was the reverse of how ranchers normally react to droughts?*

In the direst of drought conditions, instinct tells us to destock or at least spread animals as thinly as possible over the land to increase their chances of survival. I suggested instead that we combine all the herds on the brothers' four ranches into one large herd, amalgamate the ranches, and run the herd as one through the fifty-plus paddocks the amalgamation gave us. We would move the herd to a new area of fresh, unfouled ground every day or half day. I had gleaned this idea from Voisin's book, *Grass Productivity.* Voisin claimed to have noticed that when cattle move onto fresh ground, unfouled by dung and urine, they were somehow stimulated. I reasoned that this might well hold true even if the cattle only had a few leaves and twigs to eat.

As the brothers already had grazing planning experience, the replanning only took us about an hour. They were able to maximize grazing time and minimize walking distances by moving the cattle through adjacent paddocks (though they had no choice but to walk the animals, rather than graze them, on the days they moved between ranches). From then until the rains finally came, I visited the brothers often, mainly, as I said to them, "to give them a shoulder to cry on."

With no grass in sight and pathetically few leaves on the trees and ground due to the continuing armyworm infestation, the brothers' cattle fed mostly on stems and bark. We half-heartedly joked that the animals even might be eating stones. The cattle shrank down to mere skin and bones, but the losses to starvation were miraculously few. The brothers' neighbors, on the other hand—who had chosen to spread their cattle as thinly as possible over the land—experienced appalling losses. This was a lesson I would never forget.

In 1978, I had an unexpected invitation from Dr. M.E. Ensminger who ran the widely admired International Stockman's School in the United States. Dr. Ensminger was a legend in the world of animal science, sporting a long list of honors, publications, and boundless energy. He continually searched the world for new ideas and somehow had heard of my work in Africa. Would I give three lectures as part of the school's session in Phoenix, Arizona? There was no offer for financial help (after all, it was an honor to be invited!) and the sanctions on Rhodesia made personally financing the trip a daunting task, but I was not about to pass up this opportunity.

Using my forged South African passport and the one hundred US dollars I had been allowed to take out of the country, I duly arrived in Arizona. The conference was the largest I had ever attended, boasting a huge ballroom for plenary sessions and many smaller rooms for lectures. I was excited to be there although I knew no one. Time came for my first presentation, and to my disappointment it was in a small room with only six people in attendance. The next day, my second presentation drew a slightly better audience—about thirty people. My third and last presentation was scheduled for the following day. I was frankly disappointed to have drawn so little interest, especially after going to so much trouble to get there. But this time, when I reached the lecture room, people were crammed inside, the crowd overflowing into the corridor! Dr. Ensminger, noting the commotion, cancelled other lectures and moved my lecture to the main ballroom, which attendees also filled.

Immediately after my third presentation, a group of ranchers invited me to one of their rooms to talk. They wanted to know if I would advise them, but I declined. I pointed out that they were basing their request on a presentation alone—how did they know that I was anything more than a charlatan, as many experts insisted? I agreed to only assist them if they thoroughly investigated my work first. I suggested they send one or two of the most skeptical among them to visit Rhodesia.

Some months later the group selected Arizona rancher Ted Spurlock, an energetic but somewhat cynical man, to represent them, and in due course Ted arrived in Rhodesia. I shared with him as much as I possibly could, letting him visit with ranchers who endorsed my ideas as well as those who criticized them. (I arranged for Ted to spend time alone with my academic and government critics so that they could speak freely and show any evidence contradicting what I advocated.) Together, Ted and I flew and/or drove to the ranches of many of my clients, including Liebig's Ranch.

It was especially memorable taking Ted to Gumela. By then, Rhodesia was in the midst of civil conflict (the Rhodesian Bush War), and we had moved Tom's family to Bulawayo for safety reasons (our herders, friendly with the guerillas, had continued to run our cattle according to the plans Tom and I had created). By that time, we could only access the ranch in a mine-proofed armored vehicle. Prior to driving out, Ted was given instruction on how to handle the various weapons. This was all certainly a new experience for him!

At the end of Ted's visit, he returned to Arizona, reporting back favorably to the others. The group later helped me obtain the visa I would need to work with them in the United States. At the time, I had no idea just how useful that visa would prove to be.

PART VIII IRREGULAR WARFARE

CHAPTER 32

WHEN PEACEFUL TIMES FADE

The Beginning of the Rhodesian Bush War

1959–1964

Twenty people were killed on the spot—the first blood spilled in anger by the army in an impending war that would consume my life for two decades.

Working as a Territorial Sergeant during my "intensive retraining," prior to the Nyasaland Emergency.

As mentioned in earlier chapters, while still a student at Plumtree I had concluded that a military career, though a popular choice for many boys, was not for me. This was because my teenaged mind craved an action-packed life and being a game ranger offered many more life-and-death challenges than being a soldier during a time of peace.

However, my fervor to figure out why professional management failed to stop habitat destruction for both wildlife and humans was also surfacing a connection to socio-economic and political unrest. I was now seeing that poor land inevitably led to poor people, social breakdown, violence, and war. Rhodesia was not only falling apart environmentally, but politically and socially. As peaceful times began to fade, the soldier within me—nurtured by Plumtree's years of mandatory military training[1]—was ready.

[1] Up until 1953, all White Rhodesian schoolboys went through mandatory military cadet training (after 1953, this changed to mandatory three-month basic training following senior school or university). As part of my cadet training, I pushed myself to excel in every aspect, determined to be the best at everything—the most accurate shot, the best at drill and map reading, and the fastest at stripping and reassembling a Bren gun blindfolded.

The Federation[2] had both an air force and an army, the backbone of which was mainly Southern Rhodesians. The Federal Air Force was made up exclusively of pilots from the Royal Rhodesian Air Force. They were all full-time professional airmen in Southern Rhodesia with close emotional ties back to Britain's Royal Air Force, in which many Rhodesians had served during World War II. The Federation's Regular Army was modest in size, consisting primarily of a few volunteer African (Black) infantry battalions led by a core of White officers. These African battalions—the Rhodesian African Rifles (from Southern Rhodesia) and the King's African Rifles (from Northern Rhodesia and Nyasaland)—had made a name for themselves in both world wars. The bulk of the army was the Territorial Force (TF), primarily made up of the Royal Rhodesia Regiment (RRR) and including specialized units like the Armoured Car Regiment, Engineers, and Artillery. The TF was composed of conscripted White civilian soldiers who—like the US National Guard—served part-time and could be called up quickly if the need arose (as it did in World War II where Rhodesians, both Blacks and Whites, served in almost every theater of the war).

I was one of these civilian soldiers—a "Territorial"—and as mentioned in my early game department years, I was called up for intensive retraining at Llewellin Barracks in 1956, anticipating the need to support Britian during the Suez crisis, but never deployed. Thereafter I was posted to the 3rd Battalion RRR, based in Northern Rhodesia, and continued to fill a rather non-eventful role in the army. This all changed when, in 1959, the Nyasaland Emergency arose.

It started with the Nyasaland African Congress Party pleading for Hastings Banda,[3] a fierce advocate against colonialism, to return to Nyasaland after forty-two years of education and medical practice abroad. Banda acquiesced, returning in 1958 to protest throughout Nyasaland against the Federation, fearing the Federation would further deprive the rights of Blacks.

Banda spoke in English with a Chinyanja interpreter at his elbow, maybe because the cause demanded more than the teenage rhetoric he'd spoken when he left home, and maybe because roughly half of his countrymen grew up speaking other tribal languages. Nevertheless, crowds responded to his eloquence and passion, rallying behind him to topple the Federation. By February 1959, the situation had become serious enough that a state of emergency was declared in Nyasaland and the Federation's prime minister, Roy

[2] The Central African Federation, often referred to as the Federation (also known as the Federation of the Rhodesias and Nyasaland), consisted of Southern Rhodesia (renamed Rhodesia in 1964, then Zimbabwe in 1980), Northern Rhodesia (renamed Zambia in 1964), and Nyasaland (renamed Malawi in 1964). It was formed in 1953 and dissolved on December 31, 1963.

[3] Hastings Kamuzu Banda was a politician and statesman. He first led Nyasaland to its independence (and subsequent name change to Malawi) in 1964 (prior to that, he was one of many arrested during Operation Sunrise and imprisoned). He ultimately served as the leader of Malawi for thirty years (1964–1994).

Welensky,[4] initiated a secret plan, Operation Sunrise, to arrest Banda and 350 other Black activist leaders. The arrests were carried out via a burst of police raids commencing at sunrise on March 3. Welensky mobilized the 1st Battalion RRR to quell any ensuing trouble.

As I was in the 3rd Battalion RRR, I was not mobilized, but, like a good hunting dog left behind at home, I simply could not bear missing the action. Visiting my father in Salisbury at the time, I went to the King George VI Barracks and eagerly joined the queue of 1st Battalion Territorials checking in at the administrative tents.

The check-in process was well coordinated; the men's files placed in alphabetical order on trestle tables. Each man had to find his file and then report to the officer enlisting him. I knew that there would be no file for me, but I made a pretense of looking for one on the "S" table. After shuffling through the neatly organized files for some time, I reported to the officer that my file was missing.

"Impossible!" The officer responded. "We have every man's file here!"

I went back and again shuffled through the files, returning to report that yes, my file was indeed missing. The officer quipped, "If your file is not there, you are not in 1st Battalion!"

I insisted otherwise. Nearing the end of his patience with me—and not wanting to hold up the queue—the officer finally ordered me through, dismissing my file as something that would be sorted out later. Fortune continued to be on my side (or so I thought) when I was asked my rank. I responded that I was a sergeant, not knowing that we did not keep our ranks from the three months at Llewellin Barracks. Alongside uniform and rifle, I was promptly issued my three stripes.

In the Salisbury drill hall, the battalion assembled and hastily left in trucks to be flown up to Blantyre, Nyasaland in old World War II Dakotas (C-47s). Because no one knew what company I was in, I remained with about thirty other men who had volunteered to drive armored cars to Nyasaland through Mozambique, the shortest way. I quickly approached the most senior officer I could find, claiming this was a ridiculous mistake. In Nyasaland, the main language was Chinyanja. Very few if any Southern Rhodesian soldiers could speak Chinyanja; however, I had passed government exams in the language while serving in the Northern Rhodesian Game Department—and here I was being left behind! With that, I was on the next day's flight to Blantyre. There I was appointed to lead an independent platoon formed from the volunteer drivers of the armored cars.

A couple of weeks into my new sergeant role, I was summoned to appear before the colonel of the 1st Battalion. He proceeded to give me a severe dressing down. They had caught up with me—I was not in his bloody battalion, nor a bloody sergeant. I had caused administrative problems and so on. I confessed and asked if I was being sent back.

---

[4] Sir Roland "Roy" Welensky was the second and last prime minister of the Federation (serving 1956–1963). Welensky was involved in the creation of the Federation from the beginning, wishing to put the economic strengths of Northern and Southern Rhodesia to better use.

"No," said the colonel. "You will remain, and you can keep your rank as a reward for initiative!"

At the time I saw the army as an exciting calling, alongside being a game ranger. The craft of warfare consumed my interest when not thinking of game and habitat, but the causes and consequences of war not so much—that is, until March 5.

On March 5, 1959, in what is remembered as the Nkhata Bay Incident, an angry mob confronted the district commissioner on a pier in Lake Nyasa (now Lake Malawi) where prisoners captured via Operation Sunrise were being loaded on a boat for transfer to jail. The immediate degree of provocation remains in dispute, but what I heard from my men was that the district commissioner lost control, and the sergeant, leading the squad drawn from my independent platoon (and now surrounded on the jetty, their backs to the water), ordered them to fire. Twenty people were killed on the spot—the first blood spilled in anger by the army in an impending war that would consume my life for the next two decades.[5]

I might have led that squad had married men in my platoon not been prohibited from joining it. Instead, the squad was led by Sergeant Hugh Van Oppen. I had met Hugh in Kasama when he had come up to deal with troubles as part of the riot squad of the Northern Rhodesia Police. Probably, like me, he was seeking action and had somehow arrived as an unattached sergeant. Though his life was spared that day, Hugh was later killed by one of his own men while operating as a mercenary in the Belgian Congo (now Democratic Republic of the Congo).

On return after the Nyasaland Emergency, I was posted to a newly formed 4th Battalion RRR made up of mainly rural men from all over Southern Rhodesia and later mainly from Umtali (today Mutare) and the Eastern Highlands. I was commissioned and became a newly minted officer alongside several others, including renowned boxer Alan Bunnett and my friend Mike Fairlamb (who, as I've previously shared, was part of the Bulawayo Farmers' Co-op and would arrange my first lecture tour to Rhodesian farmers a few years later).

In 1960, British Prime Minister Harold McMillan delivered his famous "Wind of Change"[6] speech before the South African Parliament in Cape Town, South Africa. McMillan's speech messaged to the world that even the Conservative government of the United Kingdom would have to accept African independence. African nationalist leaders Kenneth Kaunda[7] (Northern Rhodesia) and Hastings Banda (Nyasaland) continued to rally for a dissolution of the Federation and establishment of Black majority-governed countries.

---

[5] Until the Nkhata Bay Incident, it had been sixty years since any serious violence in the region, and that was when Sir Harry Johnston defeated the local slave traders at nearby Karonga in 1894.

[6] The "Wind of Change" speech was named for its prescient warning: "The wind of change is blowing through this continent. Whether we like it or not, this growth of national consciousness is a political fact."

[7] Kenneth Kaunda, or KK, was at the forefront of the struggle for independence from British rule. In 1964, he became the first president of independent Zambia (formerly Northern Rhodesia).

Our battalion was called up frequently to quell urban riots and violence among Blacks who were at each other's throats as their political leaders squabbled and fought for dominance within the non-tribal Zimbabwe African People's Union (ZAPU)—a political party founded in 1961 by Black activists Joshua Nkomo[8] as president, Ndabaningi Sithole[9] as chairman, and Robert Mugabe[10] as publicity secretary.

Few encounters are as frightening as attempting to diffuse an angry crowd of people; however, a mob is more respectful when confronted by disciplined troops than a rabble of police, and through practice we became very adept at subduing urban riots and abating political violence with minimal harm. Riot control was a smartly executed drill in which we would first signal our readiness by fixing bayonets in unison. Then an officer would broadcast a warning over a loudspeaker. These two actions were usually adequate; if not, we would follow with tear gas. Simultaneously efforts would be made to identify principal agitators and notify snipers to target those agitators if necessary. Firing was always the last resort; seldom, and never in my experience, did we have to shoot.

In one incident the crowd got terribly unruly, and the officer in charge, while proceeding through the prescribed steps, noted a large woman as the primary agitator. It was a hot day, and she like everyone was sweating profusely as she shouted abuse at the troops and egged on her compatriots. Finally, to avoid shooting the agitator, the officer in charge ordered that tear gas canisters be lobbed in front of the mob. With much ranting and raving, the female ringleader ran out of the mob and squatted over a nearby tear gas canister, her billowing skirt enveloping the smoke, shielding the crowd from it. She squatted a moment, triumphant, and then abruptly, with an almighty scream, ran off as fast as she could. The demonstrators burst into laughter, and all was over. The woman had not known that tear gas dreadfully stings any part of the body where there is moisture!

British politicians showed little understanding of White Southern Rhodesians' fervent loyalty to Britain. Colonial administrators elsewhere were vacating Africa in haste, but we were going nowhere. We saw ourselves as Africa-born citizens who had served the Crown through two world wars, even declaring war on Germany before any

---

[8] Joshua Mqabuko Nyongolo Nkomo and others founded the National Democratic Party (NDP) in 1960, but the government banned it in 1961. Nkomo immediately founded and led the (primarily Ndebele) Zimbabwe African People's Union (ZAPU), also banned nine months later in 1962. Like other nationalist leaders at the time, he was arrested by the Smith government in 1964 and held in detention for a decade. In 1976, ZAPU merged with Mugabe's ZANU to form ZANU-PF (Patriotic Front).

[9] Ndabaningi Sithole studied teaching in the United States from 1955 to 1958. In 1963, alongside Mugabe, he founded the Zimbabwe African National Union (ZANU), a primarily Shona party that opposed the government of Rhodesia. In 1964, Prime Minister Ian Smith banned ZANU and in June of the same year Sithole was arrested and detained for ten years.

[10] Robert Gabriel Mugabe worked as a schoolteacher before joining African nationalists in calling for an independent state controlled by the Black majority. Like Nkomo and Sithole, Mugabe was arrested and held in detention (1964–1974); upon release, he established his leadership of ZANU, replacing Sithole. In 1980, he became prime minister of Zimbabwe (formerly Rhodesia). Praised as a revolutionary hero, he also later became e a dictator responsible for widespread corruption and human rights abuses.

of Britain's dominions. We did not need independence from Britain, which we in effect already had (Southern Rhodesia had been a self-governing colony since 1923). And our country was more developed than many countries in Africa, especially in terms of infrastructure, agriculture, economy, and education for both Blacks and Whites.

At this time, I would have described my enthusiasm for both the military and politics as apolitical—a naïveté which would change, but not for some years. I was not alone; many who had lived under the stable Southern Rhodesia government of Prime Minister Godfrey Huggins[11] were apolitical. Huggins and those before him had believed in a common voters roll, where all people were qualified to vote, whether they were Black, White, Asian (Indian and Pakistani), or Colored (mixed race).

Everything began to shift in December 1962, following what proved to be Rhodesia's last democratic election. A new political party, the Rhodesian Front (RF), came to power by a slender majority under the leadership of a wealthy and respected tobacco farmer, Winston Field, who was elected Southern Rhodesia's prime minister.[12] Where prior to this the government had been committed to gradual progress to Black majority rule—as enshrined in our 1961 constitution—the RF clearly intended that government would always be in the hands of Whites. This concerned many, including me. People like me referred to our new government of inexperienced racists as our "cowboy government."

The following year—in August 1963—Sithole, in conjunction with Mugabe, broke away from ZAPU, forming a tribal (Shona[13]) party, the Zimbabwe African National Union (ZANU), which only increased the violence, and the Federation began to unravel.

On December 31, 1963, the British government dissolved the Federation, granting full independence to Nyasaland and Northern Rhodesia. Both quickly transitioned to Black majority rule. (As mentioned earlier, Nyasaland, in its newly independent state, became Malawi. Northern Rhodesia, in its newly independent state, became Zambia. With there no longer a "Northern" Rhodesia to distinguish from, Southern Rhodesia became simply "Rhodesia.")

Southern Rhodesia, now Rhodesia, returned to being a self-governing colony, independent except for a few traditional practices.[14] While there wasn't a need for Rhodesia to demand immediate total independence from Britain, the racist hotheads in the new

[11] Godfrey Martin Huggins, 1st Viscount Malvern was a Rhodesian politician and physician. He was the fourth and longest-serving prime minister of Southern Rhodesia (1933–1953) and remained in office as the first prime minister of the Federation until October 1956.

[12] Winston Field founded the Rhodesian Front (RF) political party with Ian Smith. He was the seventh prime minister of Southern Rhodesia, elected in 1962, and served until he was replaced by Ian Smith in an internal coup in 1964.

[13] The Shona were likely the first Bantu-speaking people living in Rhodesia (now Zimbabwe). They are divided into five major ethnic groups: the Karanga, Ndau, Manyika, Zezuru, and Korekore.

[14] Those few traditional practices were that the British Monarch appointed our Governor General who served as the Queen's representative; the British Privy Council was the final court of appeal; and war could not be declared on any nation without Britain's approval. These practices were also still tradition with Canada, New Zealand, and Australia.

party insisted on it as Britain was pressuring Rhodesia to join Malawi and Zambia in transitioning rapidly to Black majority rule.

In April 1964, in an internal coup, the RF replaced Winston Field (who was out of the country) with a new prime minister, Ian Smith. War became all but certain under the government of Smith, whose ideology stemmed from an extreme level of racism the likes of which we had never known. Mugabe and Sithole, in addition to Nkomo and other Black nationalist leaders, were rounded up, arrested as political agitators, and held in detention for the next decade. The subsequent escalated civil conflict, coined the Rhodesian Bush War (1964–1979),[15] pitted three political forces against one another: Smith's White minority-led government versus the Nkomo-led ZAPU[16] versus the Mugabe-led ZANU.[17] The war would last more than a decade, truly upending my thinking that I was just a solider in a time of peace.

[15] Historians and others later referred to our civil war as the "Rhodesian Bush War" though we Rhodesians never did. The Rhodesian Bush War is documented as starting in 1964 (with the breakup of the Federation and the rounding up of Black political activist leaders who were put in detention) and ending in 1979 (when the British negotiated a settlement with the political leaders). For those of us who lived through it, however, most of us date the war's beginning as 1959.

[16] In the Rhodesian Bush War, ZAPU's guerrilla army was ZIPRA—the Zimbabwe People's Revolutionary Army.

[17] In the Rhodesian Bush War, ZANU's guerrilla army was ZANLA—the Zimbabwe African National Liberation Army.

CHAPTER 33

SWEARING OFF SMELLY SOCKS

Introducing the Fineries of Bushcraft to the Army

1960–1965

Digger said he had never seen or heard of one man outperforming an entire battalion.

(*left*) I'm seated, center, with my back to the camera at Llewellin Barracks; I'm in standard combat dress. (*right*) Taken on an airfield, I'm on the right in my "redesigned battle dress." Next to me is Lieutenant General Keith Coster (commander of the Rhodesian Army) and next to him is Minister of Defense Jack Howman.

Southern Rhodesia had, in its short life as a self-governing nation, governed surprisingly well, with violence, crime, and corruption almost unknown, to the extent that I had grown up in a country with an unarmed police force (British South Africa Police, or BSAP). Even as late as 1965 there was no need to arm the police. This was exemplified when that year I took a foreign photographer to document the desertification and wildlife deaths in the Tuli Circle on the Botswana border. Upon arriving at the Tuli Police Station, the photographer and I found the Shashi River in full flood, hundreds of yards wide. I asked the White policeman in charge if I could borrow his rowing boat, which lay on the river's bank. He immediately agreed but it being a Sunday and work slow, he also asked if he could join us, offering to take two prisoners along to row the boat.

Once the prisoners rowed us to the far shore, the policeman, photographer, and I walked off, leaving the policeman's .303 rifle (brought in case of a hippo attack) with the two prisoners on the boat. The photographer, puzzled, questioned our safety, "Why on earth are you both walking off into the bush, leaving the rifle and boat in the care of two prisoners?" The policeman laughed and shared that just months before, two prisoners had escaped; he had simply released all the prisoners to go and find the fugitives and offered them a reduction in sentence if they did. The prisoners soon returned with the two escapees. So, he concluded, "We are safe!" And indeed, we were—at least at that time.

Safety—and peace for that matter—were now being disrupted by a racism the likes of which Rhodesia had never known. A diminutive White sector of Rhodesia's multi-tribal, multiracial population was taking ever-increasing control of the police, army, air force, civil service, judges, newspapers, radio, and television.

While I still considered myself apolitical, I held great interest in understanding what the army's role should be in the escalating unrest. This led me to—in addition to perfecting my tracking skills—the study of irregular warfare, especially guerrilla warfare. I consumed everything I could get my hands on about the history of irregular warfare in Europe, Vietnam, Malaysia, the Philippines, and elsewhere. I studied the theories of practitioners like Che Guevara who promoted guerrilla warfare in Latin America and General Edward Lansdale who opposed guerrilla warfare in the Philippines and Vietnam. I mercilessly mined family and friends for information—from my godfather Eustace and game ranger colleague Frank Ansell (as I've mentioned early on, both my godfather and Frank served in the British XIV Army's Peacock Force, operating with Karen tribesmen behind the Japanese lines in Burma) to Provincial Agricultural Officer Bill Verboom (who had fought with the Border Scouts in Borneo against Sukarno).

At our army headquarters in Salisbury, we had a reasonably sized library where we could borrow books. I noted that very few officers read anything concerning irregular warfare. If they did, they chose material about special operations within large armies, like the Long Range Desert Group (LRDG) and Special Air Service (SAS)[1] or the United States equivalent, the Green Berets or Navy Seals. Generally speaking, guerrilla wars only break out as extensions of politics when one side has an organized, well-equipped army or police force, and the other no power or vote at all, which was indeed our situation. Yet there appeared to be little interest in learning about partisan groups who had fought to achieve political aspirations without the backing of a large army. I felt that although the politics of an anti-colonial insurgency like ours—in a time of rapidly dying colonialism—may differ in character from resistance to an invader like the original *guerrilla* harassment of Napoleon's army in Spain, similar tactical thinking applied. Conversations in the officers' mess indicated a troubling lack of awareness of what our role would or should be.

[1] The Long Range Desert Group (LRDG) was a reconnaissance and raiding unit of the British army composed mostly of New Zealanders and Rhodesians during World War II. It was augmented in July 1941 when commando officer Captain David Stirling established the Special Air Service (SAS).

My studies of guerrilla warfare had convinced me that as soldiers we were fighting a war that could only be "won" by political negotiation, not military intervention (otherwise, we would inevitably revert to squabbling between tribal and racial groups and face decades of bad governance, suppression, and corruption). I believed our role as soldiers was to limit the extent of the war, countering Black Rhodesian guerrilla gangs to buy time for Prime Minister Smith to negotiate a mutually agreeable constitution and settlement with our Black leaders. Only politicians could bring the war to a conclusion that was satisfactory to all our people. Above all, the army and police, operating amongst the people—urban and rural—had to be friends of the people (their own families among them), not affiliated with a political party, and certainly not committing atrocities, communal punishments, or maltreatment. I found myself preaching this thinking to anyone who would listen, but the response was tepid, if not simply bemused.

The only person who took my views seriously and with whom I could have intelligent discussion about them was my close friend and fellow officer, Mike Fairlamb. Mike made me realize that I was wading deep into unfamiliar waters, but I was determined to have my perspective heard.

In 1960, I had written a memorandum to our then prime minister, Roy Welensky, urging that we train our army for guerrilla warfare, including the addition of trackers. He rejected the idea. Years later when, as Sir Roy Welensky, he became a friend and political adviser to me, he said that his decision was made on the advice of his senior army officers—an understandable reaction from Regular Army officers.

But I would not give up. At every opportunity I talked with superior officers about the need for us to stop training for regular warfare and to train for guerrilla warfare instead. Although unheeded, word did get around, notably so as I learned one evening when General Jock Anderson[2] was a guest in our officers' mess. Anderson was a burly, dominating figure with a loud voice, and he could not resist, when standing close to me, to embellish on what he thought of "fucking young officers who wanted to form private armies!"

Early on (before the Federation dissolved), the officers of C Squadron[3] in the SAS—recognizing that their men might benefit from greater skills at least when lost in the bush—invited me to lecture on survival. C Squadron was stationed at Ndola in Northern Rhodesia, but while they were parachute training near Salisbury, I gave my first lecture to them

[2] General Jock Anderson commanded the 48th Gurkha Infantry Brigade with the 6th Gurkha Rifles while a Lieutenant Colonel in the Rhodesian African Rifles during the Malayan Emergency. From June 1963 to October 1964, he commanded the Rhodesian Army. In 1964, he was "forcibly" retired by Smith on the basis of age (over fifty) because he strongly opposed Smith's desire for a Unilateral Declaration of Independence (UDI) from Britain, which Smith made reality in 1965.

[3] The C Squadron of Britian's SAS (Special Air Service) was formed during the Malayan Emergency (1951–1953). A group of men from Southern Rhodesia volunteered to go to Malaya where they were known as the Malayan Scouts, which the SAS eventually renamed C (Rhodesia) Squadron, based in Northern Rhodesia. The squadron was disbanded when Rhodesia became Zimbabwe and no longer exists. However, the British army still lists 22nd SAS Regiment as four squadrons known as A, B, D, with C Rhodesia Squadron blank.

in an aircraft hangar and then later a second in the field outside Salisbury at Domboshawa. For both, I allowed no notes, considering it unwise to have anything in writing that could fall into our opponents' hands. However, Colonel D.H. Grainger later authored a slim publication, *Don't Die in the Bundu*,[4] clearly based on my talks. At least he showed me a draft before publication, agreeing to remove sensitive information that went beyond mere survival to tracking people who—unlike animals—could shoot back.

Finally, in 1962, I had the opportunity to put words into action. The army was by this time recognizing the accelerating change in Africa and began training us infantrymen in what was called COIN (counterinsurgency) operations. A four-day training exercise was staged for the 4th Battalion RRR in the Puckle Hills range west of Salisbury. We officers were informed that four gangs of "bandits" (kindly provided by the Rhodesian Light Infantry, or RLI[5]) were known to have entered the area (a size of about sixty by twenty miles). It was our battalion's task to hunt the bandits down over the course of a few days.

After a lengthy briefing, the colonel asked if we officers had any questions. I didn't hesitate to inquire, "Why are we committing an *entire* battalion to this task?" The colonel asked why I should ask such a question, and I replied, "We are not fighting to take control of an area of land but rather to locate and deal with a handful of renegades. This necessitates a section or, at most, a platoon. By committing an entire battalion to deal with so few men, we are playing into their hands."

Was I really suggesting that a hunt for four gangs of bandits in unruly bush across many miles of rugged country didn't need the entire battalion? The colonel would hear none of it, and my fellow officers had a good-natured laugh at my "stupidity."

Later that evening in the officers' mess I approached the colonel and proposed that although he had enjoyed poking fun at me, I was serious enough to challenge him. I proposed that if he would let me pursue the hunt on my own, I'd give him five pounds (about ten US dollars) for every bandit gang the battalion "kills." In turn, he would do the same for me, assuming he allowed me to operate alone. The colonel felt it not very sporting to bet on a sure thing, but we shook hands, nonetheless. An officer cadet named Rowland Davies took over my platoon and off I went with the codename "Sapphire" (a woman by the same name had, the evening before, put on a strip act for the battalion).

To shorten a long story, the battalion never did catch a bandit gang. By the end of the second day—after having operated through the previous moonlit night—I had tracked down and caught them all. The colonel sportingly paid up. Our Regular Army

---

[4] *Don't Die in the Bundu* by D.H. Grainger (Johannesburg: Howard Timmins, 1969 and 1984).

[5] The Rhodesian Light Infantry (RLI) was a unit within the Federation's army formed in 1961 at Brady Barracks in Bulawayo, Southern Rhodesia. Barely a year after its creation it was relocated to Cranborne Barracks in Salisbury and upon dissolution of the Federation remained the only all-White volunteer battalion.

training officer, Major John Essex-Clark[6]—whom we nicknamed "Digger"—clinched the matter. He said he had never seen or heard of one man outperforming an entire battalion. From then on, I got what I had been pleading for—the opportunity to start training our Regular Army in bushcraft and tracking.

Up to this point our army had zero tracking skills—all our training was in conventional warfare (and though we had native trackers to co-opt, most of them likely wouldn't have aligned with the army's motives). The low level of bushcraft skill of most troops was not only evident when I challenged the Puckle Hills training exercise, but when I was in Nyasaland. There I served alongside men who boasted an amazing array of skills, coming as they did from all walks of life. That said, their bushcraft left much to be desired, though some refuted otherwise. So, one night, a few hours after eight sentries were posted around a facility we were protecting, I stripped down to rugby shorts and bare feet and quietly visited each sentry. Four of the eight had a bayonet poking their ribs before they even knew I was there. After that, they were more modest about their bush skills, although it became increasingly difficult to sneak up on them.

Our equipment was that of conventional warfare as well, essentially unchanged from World War II. We wore bush hats with one upturned side, heavy boots, gaiters, and long trousers. When the colonel of 4th Battalion RRR posted a list of items the men had to wear and have in their backpacks, he turned to me and stated, "Savory, you're in the game department, have a look at my list and make any changes you think are needed." I jumped at the opportunity, and with a thick pencil crossed out the entire list, replacing it with one of my own.

This was the beginning of discord between the army and me, including its medical establishment. Having spent months in the bush battling poachers, I saw no need for socks (they made feet soft, smelly, and susceptible to fungus infection), underpants (they smelled after a day), nor soap (it left a recognizable scent, and we seldom had water to wash with it). Our hats looked great on parade but were quickly knocked off in thorny bush. Long trousers and clunky boots made stealth movements impossible, not to mention they were impractical in the heat. The type of ammunition we were issued—high velocity and sharp-pointed bullets—was suitable for urban areas and long-range open country, not dense bush.[7]

Thankfully for the men's sake, my "list" won over the Regular Army hierarchy, and we ended up fighting almost entirely in tennis shoes or hockey boots (the latter what

---

[6] John Essex-Clark, an Australian, was a soldier's soldier after my own heart. He had served with the Rhodesian African Rifles against insurgents in Malaya and would later fight in the Belgian Congo and Vietnam.

[7] From my hunting experience, I suggested we needed cartridges that fired round-nose, long-length bullets of moderate velocity; bullets that would not easily ricochet or change course when hitting grass stems and sticks—and would hold direction for five to fifty yards, not a thousand. We used sharp pointed short bullets designed for open country or urban warfare which changed direction when hitting even a small twig or grass stem. This is the one item from my litany of requests that the army chose not to "upgrade."

we called our thin, rubber-soled, high-top canvas footwear), rugby shorts, camouflaged smocks, and sensible caps (as shown in the photo on the right that opens this chapter).

The wild and unpopulated lower Zambezi Valley became my training ground for SAS officers and men. What started as mere survival courses I notched up to "aggressive bushcraft" courses, interweaving both tracking and guerrilla tactics. It wasn't long afterward, however, that the political climate changed dramatically.

The blatantly racist policies of Prime Minister Smith came to a head in 1965 when Smith, without a direct electoral mandate, unilaterally declared Rhodesia's independence from Britain, an act that became known as UDI.[8] Ironically, Smith's declaration of total and immediate independence effectively lost Rhodesia's independence; up until that point convention dictated that Britain could not interfere in our affairs as an independent self-governing colony. However, the moment Smith declared UDI he broke with convention and Britain promptly declared Smith's government illegal. As a result, Britian successfully campaigned for worldwide sanctions on Rhodesia.

Overnight, common sense flew out the window and dissonance descended on our White community. Most White Rhodesians viewed Britain as betraying them; not only had they been one of the first to support Britain at the start of World War II, but proportional to the White population, Southern Rhodesia had contributed a greater percentage of manpower to the war effort than Britain itself.[9] Rhodesia had committed to Britian in their hour of need, and now Britian was not reciprocating. The outpouring of emotion was further fanned into flames by Prime Minister Smith who, as a young Rhodesian, had flown a Spitfire in the Royal Air Force in North Africa and Italy, been shot down, and then escaped from behind enemy lines.

Never in the army did we talk politics, and it had always been tradition (and good manners) for officers not to discuss politics in the mess. We were a neutral, nonpolitical army; to express a political opinion was an unpardonable lapse. This proved true in 1964, when the commander of the RLI, Colonel Peter Walls,[10] had paper hats made with "RLI for UDI" written on them. When Colonel Walls distributed the hats at a Christmas dinner, Major General Sam Putterill,[11] Head of the Armed Forces, reprimanded him for such

---

[8] Rhodesia's Unilateral Declaration of Independence (UDI) was a statement announcing that Rhodesia (formerly Southern Rhodesia) now regarded itself as an independent sovereign state. This was the first unilateral break from Britain by one of its colonies since the US Declaration of Independence in 1776.

[9] In World War II, 40 percent of White men in Southern Rhodesia served in the British army, 1,720 of them as commissioned officers.

[10] George Peter Walls, who received his military education at the Royal Military Academy Sandhurst, was promoted to lieutenant colonel in 1964 and appointed commanding officer of the 1st Battalion of the RLI. He later served as Lieutenant General, commanding the Rhodesian Army from 1977 until 1980, when he served briefly as Zimbabwe's military commander and retired later that year.

[11] Major General Sam Putterill was commanding the Rhodesian Army when Prime Minister Smith declared UDI in November 1965. After warning Prime Minister Smith that he could never go along with Smith's plan to turn Rhodesia into a republic, Putterill took early retirement and became a leading light in the liberal nonracial Centre Party (a party that became irrelevant when Rhodesia became Zimbabwe in 1980).

blatant support of a political party. However, while Putterill later took early retirement in protest of UDI, Colonel Walls went on to become Head of the Armed Forces himself.

On the first call up of my 4th Battalion after UDI, the colonel commanding the battalion announced to us officers that we would be going it alone, having been betrayed by Britain. To ecstatic cheers, the officers snapped the British crowns off their badges. We would no longer be the RRR—Royal Rhodesia Regiment—but from now on simply the Rhodesia Regiment.

My enthusiasm was not equal to that of my fellow officers. I had sworn an oath of allegiance to Rhodesia and our constitution, not to Smith and his racist political party. The atmosphere was way too heated, however, so I simply pretended to snap the crown off my badge. To this day, my RRR badge remains pinned to my bush hat, crown intact, where it holds an honorable place in my study.

While I felt very alone in my reactions, unknown to me at the time, there were other far more senior army officers, air force officers, and civil servants of similar mind. The highest ranking of these men were quickly identified and removed by Smith (like General Jock Anderson). Had I even known about the torn loyalties amongst our military commanders and senior civil servants, I doubt I could have done much as my political power was insignificant at the time. With the UDI and imminent sanctions, all I could do was to work as closely as possible with the army and police to buy time for a political solution. A White minority with allegiance to Smith personally controlled the security forces, now standing in opposition to the vast majority of the country's population.

CHAPTER 34

UNCONVENTIONAL WARFARE

Leading Guerrilla Anti-Terrorist Units (GATUs)

1965–1966

The rules were simple—get to every marked place at the required time, at any cost. Critical to doing this would be outwitting, ambushing, looting, and "killing" other gangs, and above all, surviving.

SAS (Special Air Service) training in the Zambezi Valley, prior to forming Guerrilla Anti-Terrorist Units (GATUs). I'm paddling the dugout, having swum through the croc-infested river to fetch it from the other side (1963).

Knowing that violence in Rhodesia would only increase, I put my heart and soul into holding the country together through whatever I could do in my military role. I began to spend months in the army, committed to diving deep into bushcraft training and seeking approval to train for guerrilla warfare. The stresses at this point in my life were indescribable; I was simultaneously trying to continue my scientific work, support my family, manage my own ranch, and maintain my consulting practice across five countries.

Exploiting the Regular Army's favorable reaction to my "win" in tracking down bandit gangs in the Puckle Hills challenge, I presented a proposal to army headquarters asking to train what I called Guerrilla Anti-Terrorist Units (GATUs). These units would be made up of mixed-race teams from the SAS and BSAP police force. After some discussion my proposal was accepted, as was the fact that GATUs would remain top secret. I would report to General Andy Rawlins (on behalf of the army), as well as deal with Bill Crabtree, head of the BSAP Special Branch (on behalf of the police). No others were to know of our activities, although Central Intelligence (headed by Ken Flower) must have known. Originally promoted to Captain in an infantry battalion, then transferred to intelligence to free up time to do bushcraft training, I now found myself in no-man's-land. I was officially neither an intelligence nor an infantry officer in a Territorial Force unit, but on my own—a TF officer forming and training joint Regular Army SAS-police guerrilla groups answerable to General Rawlins only and otherwise invisible.

The initial training and final selection of each group of GATU recruits took place over three months in late 1965, with a weekend break at the end of each month during which I tried to see my family and keep my businesses running. The first of the GATU candidates were selected from Black and White members of the BSAP Special Branch and Criminal Investigation departments, along with officers and men from the SAS[1] (now stationed in Salisbury but including none of the men I had previously trained, all of whom had quit when offered a golden handshake retirement package at the breakup of the Federation).

The first of the new SAS training candidates showed up at my home in Bulawayo in two machine-gun-mounted "Sabre" Land Rovers.[2] The first to step out was Lieutenant Brian G. Robinson,[3] who would become a lifelong friend and comrade-in-arms, followed by a curious group of men, ready for training. Together, we travelled to the training site, on the way picking up the police candidates (some White but most of them Black). Due to the highly classified status of the GATUs, neither the men who volunteered from the SAS nor those selected from the police force knew what they were getting into.

The training site I selected was a remote, uninhabited area in the southeast lowveld of Rhodesia. I had lived there in 1959, when I oversaw the tsetse fly operations, and I knew the area well. I based the unit at my former camp homesite isolated on the Lundi River, above its junction with the Sabi.

---

[1] Upon the dissolution of the Federation the men in the SAS—many of whom I had trained—were given the choice of taking a buyout or throwing in their lot with Southern Rhodesia (now Rhodesia, to which the SAS was being relocated from Northern Rhodesia). Only thirty-one chose to go to Rhodesia, where they were stationed in Salisbury under the command of Major Dudley Coventry.

[2] The SAS used stripped-down, long-wheel-based Land Rovers with mounted machine guns that we called "Sabres."

[3] Lieutenant Brian G. Robinson was an SAS officer who later became the longest-serving commander of the SAS.

Every month I set out to train a group of approximately twenty men to operate either in the country or across the border as pseudo-gangs. The daily routine included tracking, anti-tracking, survival, and guerrilla tactics. There was one incident in which the men had good cause to curse me, though they all recovered. I was attempting to teach what I'd seen a poacher from my game ranger days do—hide from pursuers by immersing himself underwater with only his lips visible to suck in air.[4] I had tried to imitate his skill but found I could not do so without a reed to breathe through. This, too, was difficult unless the reed was the right length and diameter to enable breathing during a lengthy submersion time. To have everyone practice, I chose an isolated crocodile-free pool. Unfortunately, being a stagnant pool, it was full of schistosomes and infected every one of us with a bad dose of the dreadful bilharzia.[5]

Each intake of trainees culminated in an intense four-day challenge. For the challenge, the group was divided into gangs of four, each man given one empty water bottle and a knife. No food, water, or "luxuries" like matches or compasses were allowed to be packed, and all of us (including myself) were clothed in my "redesigned battle dress"—a shirt, rugby shorts, and canvas tennis shoes (at the start, all gangs moved out of camp some distance, then stripped naked so I could ensure no one had hidden food, water, matches, or items like catapults [slingshots]).

As if that wasn't enough to fuel desperation and conflict, one gang was given a .22 rifle with silencer, but no ammunition. The rest of the gangs were given .22 ammunition, but no rifle. Thus, no gang could get a decent meal without capturing whichever gang had the missing component. Gangs were forced to subsist on plants they had learned were edible, as well as insects and scorpions ("bush prawns"). Our frequent moves made it difficult for anyone to trap small animals or birds for food.

Each gang received a map with marked places that they had to report to at certain times. This required them to move a great deal, rather than simply lie low and survive. I deliberately selected the wildest bush for these exercises, and intentionally mapped check-in sites that would ensure gangs had to cross one another's tracks at some point.

The rules were simple—get to every marked place at the required time, at any cost. Critical to doing this would be outwitting, ambushing, looting, and "killing" other gangs, and above all, surviving. If someone could not make it, ditch the person. If someone got lost, leave them. Gangs were permitted to deal harshly with other gangs, stripping them of all belongings, including clothes. The exercise was brutal. No gang ever got to all marked places intact. At best, one man (Lieutenant Robinson) completed the entire exercise, after having ditched the rest of his gang. I was not striving for teamwork

[4] I mentioned this highly skilled poacher—who managed to elude two of my game scouts by hiding underwater—in Chapter 16.

[5] This was not my first encounter reckoning with bilharzia; for more on the disease, see Chapter 21.

in this exercise; my sole purpose was to assess the level of skill, mental strength, and determination of every GATU member.

The teams knew I was always in the area, surviving like they were, but unseen by them. I constantly tracked them, observing what each gang was doing. From each intake I selected the best candidates to be instructors to assist with the next intake. It was the best way I'd found to train men rapidly.

On actual operations we used enemy arms and clothing, as the enemy's gangs did, and blackened our faces and hands with a dye. To avoid giving ourselves away, there could be no deference toward a gang member who was actually a White officer dyed black. To reinforce this, I got the Black recruits to give White officers nicknames and refer to them by their nickname, regardless of rank.

Africans are well known for giving nicknames that are very apt and descriptive of a White man's features, character, or behavior. They nicknamed me *Chimurenga* which was the Shona name for the various rebellions against Cecil Rhodes' British South Africa Company rule in Rhodesia's short history. When I asked why they had chosen that name, they said, "You are constantly teaching us how to wage guerrilla warfare for peaceful ends, but no man can know so much about war unless he is himself warlike."

GATUs were still a far cry from having higher officials understand guerrilla warfare, but it was a hopeful next step. Complicating the situation was my worry that we were in danger of becoming a political tool if Prime Minister Smith did not quickly negotiate a better future with our Black countrymen. One evening in our GATU camp, when army and police officers were sitting in the bush discussing the war, I tested the water by saying I was not sure who the enemy was because every man involved on both sides was a Rhodesian. That went down like a lead balloon, and I scrupulously did not bring it up again. Tragically, my concern turned out to be all too real; our military successes ultimately only increased Prime Minister Smith's intransigence.

Mercenaries had increasingly volunteered to join our army. Some were veterans of the Vietnam war, others were veterans of various African conflicts, and many of them rogue soldiers of fortune who saw little wrong in brutalizing civilians. When Africans reported to me that the White soldiers who had burnt their villages could not even speak the common pidgen language Chilapalapa,[6] that confirmed these were mercenaries. In a guerrilla war especially, brutalizing civilians only increases their support of the guerrilla fighters, as the German army learned in World War II when their brutal treatment of civilians behind the lines turned the entire Russian population into resistance fighters.

The army could not, however, just blame the mercenaries. Many of our own police and army officers were emboldened to engage in immoral actions sure to cost support

[6] Every Rhodesian-born child, Black or White, knew the common pidgin lingo, *Chilapalapa*, even if clueless about our various tribal languages.

among our Black population. They laid minefields, assassinated and abducted people, distributed poisoned clothing, forced families into enclosed settlements that amounted to concentration camps, and burned their villages. I took no pleasure in being the first to officially expose an atrocity committed by two South African policemen who were operating in Rhodesia. When a mother could not—or would not—answer their questions, they ruthlessly cut the throat of her baby. To my dismay, the whole affair was hushed up by both the South African and Smith governments. I then arranged for the matter to be raised in Parliament by its African members, giving Minister of Defense Jack Howman an opportunity to condemn the actions, which he shockingly chose not to do.

When I had completed selecting and training the first GATUs, we tested going operational on our northern border in an area first cleared of other troops. While setting up base camp in the escarpment hills of the Zambezi, Crabtree and General Rawlins sent me a new regular officer who I had not trained, Captain Spike Powell. Spike had only recently joined our army, having been involved with pseudo-gangs in Kenya. It was not long before Spike raised the question of how we should handle things if accidentally discovered with our blackened Whites and enemy weapons. He proposed that if this were to happen, we kill whoever discovered us, then disguise the murder as an atrocity committed by ZAPU or ZANU "terrorists." He pointed out the obvious: the government and army would deny any involvement or protection. Spike's proposal was totally unacceptable to me and against every GATU principle. Rather than have a showdown with Spike (being that he was the same rank as me but a Regular Army officer and me a Territorial officer), I insisted that we put this "what if" to rest until it might occur, determined that I would prevent it. Thankfully, we were not discovered.

What we did not foresee was that it would not take long for challenges to rise in the GATUs because the units required such a close working relationship between army and police, yet the command and control of the units remained undefined. After a series of high-level discussions to which I was not privy, the decision was made to dissolve the GATUs. The police members went on to form PATUs (Police Anti-Terrorist Units) that served well but bore no relation to the original GATU concept because they were merely rural police reserve units commonly operating from Land Rovers. The SAS members went it alone but continued training trackers through the formation of the Tracking Wing of the School of Infantry at Kariba. And as for me, the no-man's-land I had been in was now truly one man's.

CHAPTER 35

BUT FOR THE TASTE OF TARZAN BARS

Leading the Tracker Combat Unit (TCU)

1967–1973

Twenty men in single file marched past me. I counted each one, breathless as they passed, ready to jump up and open fire should there be any hesitation in their steps.

Tracker Combat Unit (TCU) men on a tracking exercise. Their bandoliers, which I designed, contained pouches that could hold a week's supply of food.

White game rangers—as well as professional hunters and their safari clients—traditionally depended on native trackers. Few White rangers or professional hunters ever hunted on their own, and fewer still did their own tracking. It was only when I started hunting man-eating lions as a game ranger that I became determined to perfect my tracking skills. This was due in part because native trackers did not share my enthusiasm for closing in on predators in dense bush; as we would gain proximity to a lion, fear would naturally escalate, and often trackers would suddenly find the tracks "harder to see" and then, mysteriously lose the spoor. After a few such episodes I realized

I simply had to become a tracker. Being somewhat fanatical and not shy to tap into the wisdom of older, more experienced native experts, I spent endless hours tracking, determined to be the best tracker and the most skilled in all aspects of bushcraft.

It was a life-changing education. I learned that there is far more to tracking than simply following tracks. It is essential to get into the mind of your quarry—whether animal or man—to understand why they're doing what they're doing, what they're thinking, and what they'll likely do next. This mind reading is gradually built based on instinct and countless clues gleaned from the slightest disturbances in soil and plants; the tracks of insects over the spoor; broken spider webs; wind, dew, and the angle of the sun; the surrounding noises of birds, antelope, and insects; sudden silences; and more.

With the dissolution of the GATUs, I predicted a need for a unit focused on internal intelligence gathering and tracking guerrilla fighters crossing into Rhodesia. I quickly pivoted to requesting permission to recruit soldiers from the Territorial Force (TF) for this unit, and hence the formation of a Tracker Combat Unit (TCU). I was given recruiting priority throughout the TF battalions, and within a year the TCU carried out their first successful operation.

The original members of the TCU were volunteers who came mostly from the very company I had helped create for game ranching—Wildlife Utilization Services Pvt. Ltd. (WUS)—and most of the rest from the Southern Rhodesian Game Department. My first task was to make trackers of these men (all of whom had previously relied on native trackers) and then teach them the various tactics and tracking formations that made it difficult for the enemy to ambush trackers (not to mention difficult for trackers themselves to be tracked!).

A tracker team was composed of four men. The man tracking on the spoor—the "tracker"—was followed by the tracker team controller (in command of the entire team) who also communicated via radio with follow-up troops when we had any. The remaining two trackers, known as flankers, usually walked far out to the sides and far ahead of the tracker who was on the spoor. The flankers' main task was to detect ambushes so the tracker could safely concentrate. If a flanker picked up the deviating track, he signaled to the others and immediately became the tracker. The former tracker would then swing out to replace that flanker. Only in extremely dense bush, where noise made by flankers would outweigh the advantage of their forward positions, would one or both flankers temporally pull in behind the controller.

As silence was our greatest security and advantage, I was strict about noise. Only empty or full water bottles were allowed. My reasons were twofold: firstly, a half-full bottle made noise. Secondly, as we were often thirsty and short of water, it was demoralizing and distracting to hear water sloshing. Nothing was permitted to project from the side of a man's body—not even a water bottle (note that in the photo opening this chapter, water bottles and pouches are at the back and never to the side). Leg coverings were not allowed, no matter how thorny the bush. To communicate we used a combination of hand signals, head nods, and "silent" dog training whistles tuned daily to ensure the pitch used was audible to all members (giving us the ability to project a high-pitched insect sound that was nearly inaudible to our enemy, who in any case expected us to signal with bird or wildlife noises).

When we had support troops following us, we had to keep them some way back—they were so noisy—and forbid them to use radios unless we called. Having the comfort of support troops, who never seemed to adhere to our code of silence, was not really comforting if they could only support us after we'd fallen into an ambush.

The typical army webbing that held a man's kit was much too noisy in thorny bush, so I designed TCU-specific bandoliers that a man could wear across either side of his chest The bandoliers were complete with pouches where food could be safely (and silently!) stored. One bandolier of food could last for a week, supplemented by bush foods. Each man filled his own bandoliers with food he chose.

The convenience of these bandoliers led to me making a foolish mistake on one occasion. I was on an external operation (on my own, with the SAS but without the TCU) in Mozambique and did not have my bandoliers with me. I decided to rely on new rations produced in South Africa that the SAS were testing. We called these rations "Tarzan bars" as each bar was allegedly a fully balanced and filling meal. Where others took normal rations and a couple of bars to test, I foolishly assumed they would be good, packed nothing but Tarzan bars into my ammunition pouches, and went on my way.

I had not banked on the bars being inedible. After hours in the sun, they exuded oil and reeked; I literally could not hold them down and had the immediate urge to throw up whenever I smelled or saw them. For some days I had a prisoner attached to me with a piece of parachute cord and shared my water and food with him, as he was famished. After half a Tarzan bar, he also spat the rest out, apparently preferring to die. I ended up having little option but to scrounge the army rations from the SAS troops to keep my prisoner and me alive. Later their commander, by then Major Brian Robinson, confided that this prompted his troops to grieve, "Of course Allan carries less kit—and insists we all should—because he feeds off the rations we carry!"

I reasoned that our most likely losses of trackers would be due to mines or ambushes. I taught trackers to be aware of bush sounds and their meaning. They learned to recognize and interpret the significance of animal and bird sounds, especially the first chirp of the honeyguide bird[1]—likely indicating a human was very near—and to take note as soon

[1] Honeyguide birds, also known as indicator birds, are best known for their interaction with humans. They will deliberately lead honey-hunting humans directly to bee colonies so that they can feast on the grubs and beeswax that are left behind when the humans remove the honey.

as a kudu barked, an oxpecker[2] chirped, or beetles, frogs, and birds went silent. Because we could only be mined or ambushed where we were expected, we always trained to never go where expected. I did not allow men to return to the same camp, not even if equipment had been left behind. Should any man ever be captured and forced to lead his captors to us, he would warn us via his movements. Should he be compromised, he would have his weapon in his left hand. No weapon in hand, or holding his rifle with the wrong hand, was all we would need to see to be tipped off that something wasn't right. All simple, easy, and workable.

To ensure success, teams constantly tracked each other in simulations. One team would take off into the bush—leaving as few tracks as possible—and then hide as the "enemy" in ambush. The other team would be tasked with locating the enemy and surprising them first; if they failed to do so, the enemy would shoot at them. Initially, I had the ambushing team fire live rounds at the approaching trackers, ensuring they only shot close to their feet. But eventually everyone ignored live rounds, knowing they would not be harmed. So, I replaced live rounds with powerful catapults. The stones shot from these catapults came fast and hard enough that they seriously hurt any person they hit; they shattered when hitting the magazine of a rifle. The threat of being shot by a catapulting stone kept trackers alert for hours, no matter how hot or tired. Amusingly, the ambushing team took to striving to shoot me if they could (as coach, I was always walking alongside the tracker pointing out signs missed). In the end our sense of hearing (and speed of responsiveness) became so acute that we were reacting to the sound of a fired catapult faster than the stone could reach us.

Aside from ambushes, getting lost in the bush was always a danger. To ease this fear, everyone memorized the major drainage pattern of the area they operated within. Thereafter, if they were ever lost, they simply walked downhill, which would always lead to a river and eventually some sign of human presence, most likely a road. Only over vast areas of flat ground did this not work.

I also tested each man's directional instincts, as this varies a great deal between people. Each man was blindfolded and led around on an overcast day in dense bush. Once the blindfold was removed, the man would then be tasked with pointing out the directions to camp, where we had left the vehicle, and other features now out of sight. Some pointed all over the place, while a few were amazingly consistent and accurate (I also took the test, but my directional skill was only average).

Trackers were also taught anti-tracking—ensuring they left almost no sign of their presence. I had the army collaborate with a local company to produce custom canvas hockey boots for the TCU with soles that left nearly no print. These I designed from

---

[2] An oxpecker is a small African bird known for its habit of perching on large animals like zebra to feed on ticks and other parasites.

observing elephant feet, which are superior to humans' bare feet, leaving remarkably little sign considering the weight and size of the animal.

With experience, the men also learned that, if you do need to shake people off, it is not only a matter of being trackless, but also of moving faster than your trackers. Most inexperienced men would waste time trying to avoid leaving tracks or trying to cover up their tracks. The most difficult to track is an experienced tracker who moves fast across hard-to-track ground, preferably in daylight as night movements tend to leave more signs. For this reason, when we had to steal water from a source close to an enemy camp, I would sneak in at midday.

Apart from being silent, swift, and skilled trackers, everyone needed to be above average at instant shooting in dense bush. To achieve this in training camp, I coached the men to focus on two methods: shooting with both eyes open and using sound to help determine the target.

For the first skill, I provided at my cost unlimited ammunition and my silenced .22 rifle with the sights blocked off between the foresight and the muzzle. One eye could line up the sights but not see the target while the other eye could only see the target. Only by shooting with both eyes wide open could the swinging target be hit. I encouraged the men to fire at the targets, competing as much as they liked in spare moments.

To hone the second skill, we trained to hit invisible targets in the bush, each with a rattle attached. The rattles were tied to a string that a man would pull (from a safe place) as the shooter walked by. This required the shooter to identify the target via sound only and hit it, often on his right (which is the most difficult side to shoot fast from for right-handed men).

I had acquired my shooting methods from my godfather Eustace[3] and working with experienced game hunters—all of whom had methods vastly superior to the ones used in army training. I confirmed this when four TF battalions, training jointly, gave me the opportunity to partake in their jungle range—a winding bush track with half-hidden camouflaged targets and others that popped up unexpectedly. To demonstrate that shooting instinctively at short range with both eyes open did not require sights, I removed the sights on my rifle. I then walked through the course, hitting every target under their record time.

Sometime later, when the army was teaching the British method of night firing, I again demonstrated the success of instinctive, two-eyes-open shooting. A squad of us lay on the rifle range one night and fired ten shots each, fifty yards from vaguely visible human silhouette targets. I chose to ignore what was being instructed; I lowered my head

[3] My godfather Eustace Poles in turn learned his shooting methods from W.D.M. Bell, better known as Karamojo Bell, after the Karamoja region in Uganda, which Bell travelled extensively as an adventurer and big game hunter. Bell was later a decorated fighter pilot, sailor, writer, and painter.

to the ground until I could pinpoint the position of my target by its silhouette, then raised my head, keeping both eyes open, and fired ten rounds fast. Upon inspection—the instructor shining his flashlight on the human silhouettes—each man was praised for the three or four holes in his target. When we arrived at my target, there were ten holes, neatly positioned in the silhouette's chest.

In most situations the TCU did not do sentry duty. Being on sentry duty prevented men from getting enough sleep, and within our small teams it was critical that each man be at peak performance during the day. What we did instead was prearrange a rendezvous point for the next morning and then simply scatter at dark. We usually slept hidden in the bush in pairs to help prevent a snoring man from exposing us. On one occasion I found our rivals were doing the same as I nearly walked into a sleeping enemy curled in the bush in the pitch dark. I could smell him but not see him (and he no doubt could smell me as we all rarely bathed). I stopped in my tracks, standing deathly still, scouring the darkness. After what seemed like forever, the man jumped up and bolted into the darkness from only a few yards away.

While on this same operation I had a most frightening experience. I had gone out on my own, searching for the exact boundaries of an enemy camp we had located, intending to call in an airstrike at dawn. At twilight I was carefully working along one of their entry paths when I suddenly heard men approaching behind me. I barely had time to conceal myself on one side of the path—quickly lying down, turning my face away, and hiding my rifle with my body. Thankfully, my camouflage jacket was so weathered and faded that it blended seamlessly with the short, dry grass. Twenty men in single file marched past me. I counted each one, breathless as they passed, ready to jump up and open fire should there be any hesitation in their steps. Thankfully, all were right-handed, each carrying their weapon casually on their right shoulder, blocking their view to where I lay. Once the last man passed, I intended to tag onto the end of them but remained paralyzed in place, my heart thumping with fear.

While our commanders could see the need for special forces fighting behind an enemy army's lines, they could not see that we had no enemy army or lines, but Black political parties forced by Prime Minister Smith's government to operate from neighboring territories with meager guerrilla forces inside Rhodesia. We were, at no point, fighting a foreign enemy, but rather our own people. We had to project ourselves as defenders of a civil order, minimizing violence and allowing time for political negotiation and resolution—a naïve hope.

As officers in the army, we had already departed seriously from this stance, having disavowed our oaths of allegiance to the constitution in favor of support for a political party that had forsaken any form of democracy by taking control of the media, replacing impartial judges with political ones, and removing army commanders who were not in support of these moves. Defenders of civil order was a hopeless illusion, morally, physically, and of course politically.

In the TCU, as we all were civilian soldiers, I encouraged political discussion, which flowed more freely and frequently than in the Regular Army. I stressed that—despite political opinions—I would not tolerate disrespectful treatment of villagers or prisoners. Doing so would result in the offender's immediate removal and a return to his original unit. This amiable code of conduct proved worthwhile on more than one occasion.

Once, as I gave orders to pull out of an enemy camp in Mozambique, an SAS corporal asked if he could set a booby trap. I responded, "Can you guarantee that your trap will not kill or maim a woman or child?" The man could not. So, no booby trap. Instead, I wrote a note to the enemy commander, assuring him that his camp was free of booby traps. I also commended the enemy on the camp's cleanliness and thanked him for his hospitality in his absence. I then signed the note and pegged it to a reed door with one of their bayonets. A few months later, my cook in Salisbury informed me that two young men were at the back door, wanting to see me. When I greeted them, they handed me a pumpkin with a thank-you note attached from "the boys in the bush."

I asked, "How did you find me?"

They responded, "We got your address from a phonebook at a store we raided."

I accepted their pumpkin, and we parted feeling a little more human on both sides, no doubt.

Years later, in 1973, I met with Lieutenant General Peter Walls, who by this time was Head of the Armed Forces. I requested that he relieve me of my command of the TCU, for at that point I was actively in politics and publicly opposing his handling of the war (of which I go into more in the following chapters).[4] Walls agreed to relieve me of my command.

That same year the TCU was subsequently absorbed into a new regiment, the Selous Scouts. Interestingly, no mention of the TCU appears in the official history of the Selous Scouts[5]—perhaps fortunate as the scouts themselves, although having many exceptionally courageous Black and White troops, went on to behave in a manner I would never have allowed, earning a reputation for deviousness, ivory poaching, infighting, and political intrigue.

I later received notification in the post that I was awarded a Military Forces' Commendation, Non-Operational. While an MFC had more meaning, an MFC Non-

[4] I liked Walls; my request to be relieved of the TCU command was essential but not personal. I felt honor bound to inform him that I had no confidence in the prime minister or himself to win the war—and would be publicly critical in coming months. Walls too had gone to Plumtree, then to the Royal Military Academy Sandhurst, and was commissioned in the Black Watch (an infantry battalion of the Royal Regiment of Scotland) before returning to Rhodesia. His son, George, and my son, Jim, were at Plumtree together, and on top of that his brother (also in the army) and brother's wife were amongst my sister Ann's best friends. Nevertheless, I was compelled to speak out.

[5] *Selous Scouts: Top Secret War* by Ron Reid Daly and Peter Stiff (Johannesburg: Galago Publishing, 1982). Although considered the official history of the Selous Scouts, the authors were apparently unaware of the early GATU pseudo-gangs, nor do they mention the TCU, which the scouts sought to replicate.

Operational was the lowest award the army gave, and it was—I felt, as did others—meant to be an insult. That evening, while drinking with friends in the SAS officers' mess, we gave a toast to "My Fucking Contribution" and had a good laugh. I took the citation to Walls and after thanking him for the recognition, told him I was declining the honor. He insisted that officers could not refuse awards, and I said I felt it was an insult and preferred to leave with my pride, dignity, and memories. I crumpled up the citation, put it on his desk, saluted him, and left. It was for me a sad ending to many wonderful years and comradeship in the army—an army that had fulfilled its duty through the Boer War and two World Wars, only to be betrayed by politicians and servile commanders who made an unnecessary war inevitable and unwinnable, costing tens of thousands of lives.

No epilogue describes our war better than the one published seven years later by Rhodesia's last Chief of Intelligence, Ken Flower, who said, "... a war of counterinsurgency is twenty percent military and eighty percent political—which implies that greater force is needed against politicians to settle a dispute, than against 'terrorists' in the field, or government will shelve its responsibilities and the military ignore theirs."[6]

[6] *Serving Secretly: Rhodesia's CIO Chief on Record* (Alberton, South Africa: Galago Publishing [Pty] Ltd., 1987), p. 279.

CHAPTER 36

THE ORIGIN OF SCIENCE

What My Army Days Taught Me About the Environment

1959–1973

One morning I checked the oil sumps on the SAS vehicles and was stunned at what I found.

Radioing my location from a captured enemy camp in Mozambique on my last operation (1978).

Through my active army years, I tried to keep my personal life going as normally as I could but fell short. In 1968 Shirley and I had a second son, Hugh. A bright little fellow, Hugh only lived six months before contracting encephalitis from a mosquito bite. I was regrettably not there when he died. The army radioed me the message. Upon receipt, I walked into the bush, cried my heart out until dawn, then carried on.

Though my twenty years of active involvement in the army did not end as I'd hoped, I learned a great deal. My learnings were invaluable to the bigger purpose to which I've devoted my life. The many days and nights I spent tracking down opponents, with whom I had more empathy than enmity, shaped me to better understand the continuing degradation of the landscapes over which we fought.

In teaching others how to track I had myself become an even more methodical tracker. I learned to observe at an extreme level of detail using not just sight, but sound and smell—as my life depended on it. In turn, this made me a better scientist, for the core of science is understanding nature through observation, deduction, interpretation, discussion, and—where applicable—experimentation. These aspects are embedded in the art of tracking. I would later learn that at least one author, Louis Liebenburg, put forth a convincing theory that science's origins were in fact in tracking.[1]

During many a long night lying sleeplessly in the bush—often cold and wet as building a fire would have made us easy targets—I had time to think. Why was tracking today harder than the past few days? Why did the ease of tracking vary depending on the use of the land—be it a game reserve, commercial farm, or communal land—even though it was on the same soil in the same vegetation? Why were we so easily able to spot the slightest disturbance to the soil one day and not another? Why were there no insect tracks or fine sand to aid in interpreting the age of the tracks? Why were tracks sometimes visible right in front of us and at other times only twenty yards ahead—and only then if we followed a line of clues?

The answers, I realized, depended on the complexity (or simplicity!) of the biological communities (plants, animals, birds, insects) we were operating within. The top millimeter of soil usually gave clues to the overall complexity. If the soil was soft and covered with a mulch of dead plants (litter), the life within it would be more complex as would life above ground, and—as there was seldom a footprint—tracking would have to be from what I referred to as "aerial signs of direction"—the color of grass, bending of stems, absence of dew, and disruption of spider webs. If the soil was hard and bare, the life within it would be simpler and sparse, and so would the life above ground. Our tracking would follow irregularities on the soil surface including human footprints as opposed to a litany of nature's clues above the soil. Whether a man, moving in daylight, put his foot alongside or on top of a plant told a lot about his mental state.

These factors made for different levels of tracking over similar landscapes. I realized that the difference was not due to the land itself but to what it was being used for, which in turn dictated its management. If the land was frequently burned—as national parks and commercial ranches often were—it was easy to track because the ground was bare between the plants. Communal lands, on the other hand, were at times more difficult to track upon. These lands were less often burned and had fewer animals (though always present); plants were spaced closer together; upright bunch grasses had succumbed to overgrazing; and often these grasses had been replaced by thick mats of runner grasses.

When I was doing GATU training, one morning I checked the oil sumps on the SAS vehicles and was stunned at what I found. Years prior when servicing my Land Rover in the bush, the brass oil sump plug had been difficult to remove with the spanner slipping off repeatedly because the plug had been so worn down by grass. Now—after three months of use—the sump plugs were unworn and covered in old oil caked with

[1] Louis Liebenburg's book is *The Art of Tracking: The Origin of Science* (Johannesburg: New Africa Press, 1990).

dust. The dense perennial grass in the middle of bush tracks had been replaced by sparse annuals and bare ground, no doubt in part because of the days when we thought that the way to eradicate the tsetse fly was to shoot out the game.

I had an ongoing mental filing system for these "clues," and often felt that in the army, I was engaged in "science" at a depth far more than most ecologists or fellow scientists had ever been. While my army years were a lost cause politically, what I learned through tracking gave me an education no university could have provided, nor any scientist had enjoyed. I was now able to see the land anew and to better assess its condition, and that in turn empowered me to persist in understanding the cause and remedy of the desertification that had plagued us for millennia.

PART IX PARLIAMENT

CHAPTER 37

ONE MAN, ONE VOTE, ONE TIME

Deciphering Rhodesia's Messy Situation

Early 1960s

Developing a modern state out of this demographic kaleidoscope was clearly not going to happen under Smith's racist agenda.

Prime Minister Ian Smith, surrounded by his cabinet members, signing the Unilateral Declaration of Independence (UDI) from Britain (1965). *(Photo entered public domain when Rhodesia became Zimbabwe.)*

Prussian philosopher General Carl von Clausewitz, who wrote from his experience in the Napoleonic Wars, breaks down the horrendous totality of war into elements that I've found helpful in understanding Rhodesia's situation in the 1960s. The totality of any war, he said, is a blend of three factors: The first and most subject to change is *politics*—the realm of governments. The second is *emotion*—a mass of people must be riled up enough to risk death and kill each other. This is the realm of ideologues, propagandists, demagogues, and zealots. Third is the challenge of the *game* itself. This includes weapons, military strategies, combat, the risks involved, and the camaraderie amongst soldiers in battle. All of this can be difficult to ever understand if one has not experienced it. The three factors profoundly touched my life, although I couldn't have defined them so clearly at the time.[1]

In Rhodesia, as both Blacks and Whites strived to finance and develop a nation state, it was always our belief that Parliament and civil service would eventually be led by—and composed of—the Black majority (a belief, at least, until 1965). Economic and social development would inevitably increase the number of Blacks qualified to vote as more became literate and educated. This was also the belief of Rhodesia's founder, Cecil Rhodes, who made his views known in a call for union delivered to the Cape Parliament on July 27, 1894:

> *I am working, I tell you, not only for the union of the country, but for a union of races, and that will come once the principle of equal rights is accepted for every civilised man south of the Zambezi. Now you ask what I mean by a civilised man. Well, I mean by civilised man any man, quite irrespective of colour, who can write his name, place of residence and occupation, and who is a worker or possessed of some property.*

It was by no means a simple situation. Although the demand side of the ancient slave trade in the Americas had been eliminated by Britain and America approximately 200 years prior, both the demand and supply side of this terrible trade in Africa had not. It was a mere forty years before my birth that capturing and trading slaves still transpired in Southern Rhodesia. Only with the defeat of King Lobengula[2] in 1894 by a small band of pioneers—and the subsequent release of thousands of mostly "Amasvina"[3] slaves—was slavery ended in Southern Rhodesia. This coincided with Sir Harry Johnston's defeat of the Yao slave-trading tribe in Nyasaland, which ended the trade in all of Southern Africa. Courts were then established—a novelty because up to that point, guilt was determined by the accuser, on occasion by whether the accused's hands blistered when retrieving a stone from the bottom of a pot of boiling water.

In my childhood years, I never placed much thought on why our government patronized most of its population. Millions of Blacks were emerging from centuries of intertribal warring, and I assumed our leaders were doing their best to develop a nation as fast as they could. Because Southern Rhodesia was a self-financed colony, our resources for

---

[1] For more information on Clausewitz and this theory of war, see the new and revised edition of *On War* by Carl von Clausewitz (trans. Col. J.J. Graham)(London: Kegan Paul, Trench, Trubner & Co., 1918). See especially Volume 1, Chapter 1, "What Is War?"

[2] Lobengula Khumalo reigned from 1868 to 1894. He was the second and last official king of the Northern Ndebele people (historically called Matabele in English).

[3] Amasvina, meaning "the dirty ones," was an insulting term used by the Ndebele people towards the Shona people they enslaved.

development were minimal. Until I left high school, all roads were unpaved earth except our main roads. But even then, to call our main roads *paved* was questionable as they consisted of two narrow, tarred strips the width of a vehicle's wheels. Our bridges were low-level concrete drifts that flooded when it rained. When traveling by vehicle in the rainy season, one had to drive slowly and carry blankets, food, and a pack of cards—expecting to be held up for weeks at a time. Southern Rhodesia could not afford a university, but our government and churches worked together to establish nine boys' schools, an equal number of girls' schools, and a couple of co-ed schools for we "Europeans." Government education of Black Rhodesians had been woefully inadequate compared to need (which was seen as not a deliberate intent, but a financial reality), yet the country still managed to turn out a growing number of Black graduates, some of whom I was to serve under.

I grew up under Prime Minister Huggins, the Commonwealth's longest-serving prime minister, and his successor, Garfield Todd,[4] who was a progressive clergyman before becoming prime minister of Southern Rhodesia in 1953 (when Huggins became prime minister of the Federation). Though I don't ever recall discussing politics with my friends as a child, from family conversations I knew Huggins was a respected leader, and when Todd—who was a family friend—became prime minister, we knew he was doing everything in his power to advance "Africans" into government. Todd had maintained open relationships with Black nationalist leaders, including Robert Mugabe and Joshua Nkomo. Even though his government toppled when he tried to expand the franchise to expedite Black political advancement, Todd still retained significant support among Black *and* White Rhodesians.

By the time I was an adult and working in both the Northern and Southern Rhodesian Game departments, we freely touched on racial differences, bantering in a manner that seemed healthy compared to what came later. On especially hot days I'd say to my Black colleagues, "You're lucky to have black skin that doesn't sunburn like mine!" In evenings around the campfire, we'd chuckle about how White women lay in the sun to make their skin darker, while Black women bought creams to make their skin lighter. It was clearly racist to compare a Black to a baboon (or any other ape), however two of my game scouts light-heartedly compared me to a baboon one afternoon while sitting out a rainstorm. They laughed when I squatted and clasped my hands over my head to fend off the rain. I didn't get the joke until they pointed out that the water running down my hairy arms revealed my kinship with baboons: a White man's hair runs down on his outer arms and up on the inner sides, just as a baboon's does.

As previously mentioned, in 1962, a conservative political party, the Rhodesian Front (RF), led by Winston Field, orchestrated a blatantly racist election campaign,

---

[4] Sir Reginald Stephen Garfield Todd was a liberal prime minster of Southern Rhodesia from 1953 to 1958. He later became an opponent of White minority rule in Rhodesia.

fanning the fears of the White population. I remember posters plastered everywhere showing just the legs of Black schoolboys clad in khaki shorts next to the legs of White schoolgirls clad in skirts with the headline, *Is This What You Want?* To the surprise of most Rhodesians, the party won. Field became prime minister, and the RF gained a majority of twenty seats in a fair and democratic election (the last in the country to this day, some sixty years later).

From the memoirs of Ken Flower,[5] who headed Rhodesian intelligence, we learned of two underreported factors that contributed to the RF's win. First, the retiring head of police (BSAP), Basil Spurling, wrongfully engaged in politics, speaking in favor of the RF to thousands of farmer police reservists and their families. Second, ZAPU's leader Joshua Nkomo—then rising in popularity among Black nationalists—asked Black Rhodesians (who then had a vote) to boycott the elections.

Field was prime minister for only a short time (December 1962 to April 1964) before he was replaced by Ian Smith in an internal coup, the details of which we did not know until years later. Again, from Ken Flower we learned that a Unilateral Declaration of Independence (UDI) from Britain had been under consideration within the RF, but all the advisors opposed it, including Field himself. Infuriated, Smith and his followers from within the RF decided to oust Field while he was away in Britian. Smith was suddenly leading Rhodesia (named so after the Federation dissolved on the last day of 1963 and Southern Rhodesia dropped "southern" from its name), determined to suppress Black majority rule once and for all.

Just four years earlier, stirred by British Prime Minister Harold Macmillan's "Wind of Change" speech, Black Rhodesian nationalist leaders had begun vocalizing a demand for "One Man, One Vote." This sounded fair to the Commonwealth and the world, assuming it would result in reasonable Black rule.

However, many Rhodesians—including me—knew that in practice One Man, One Vote literally meant immediate rule by the winning party—and the end of democracy. It would be One Man, One Vote, *One Time.* The winning party would hold power at all costs, a syndrome which still plagues Rhodesia (now Zimbabwe) and much of Africa to this day.[6]

Sure enough, once Smith was at the helm, he and his party turned deaf ears to Britain's insistence that Rhodesia stay steadfast on the path to Black majority rule. Smith had the backing of many White Rhodesians who were terrified by the stories they'd heard from

---

[5] Ken Flower was appointed Deputy Commissioner of the BSAP (British South Africa Police) in March 1961 and subsequently became head of Rhodesia's and later Zimbabwe's Central Intelligence organization, serving under four prime ministers, including Mugabe. His memoir is entitled, *Serving Secretly: Rhodesia's CIO Chief on Record* (Alberton, South Africa: Galago Publishing [Pty] Ltd., 1987).

[6] *One Man, One Vote, One Time*—messaging that the winning party would hold power at all costs—was certainly the case for Northern Rhodesia and Nyasaland. When the Federation dissolved in December 1963, Hastings Banda held power in Nyasaland (renamed Malawi) for three decades (1964 to 1994); similarly, Kenneth Kaunda held power in Northern Rhodesia (renamed Zambia) until 1991.

Belgian Congo refugees flooding into Rhodesia after the violent and chaotic collapse of their country during independence. Smith and the RF were adamant that such a collapse would never occur in Rhodesia; they reinforced this by declaring immediate and unilateral independence from Britain.

The most prevalent fear of White Rhodesians, however, was that our country was in danger of being completely overwhelmed by Black "communists." Smith exploited that fear to further justify his actions. By 1965 he had not only removed higher military commanders opposed to UDI, but he had also taken effective control of everything essential to free and fair democratic elections—judges, army, air force, senior police officers, and the Rhodesian Broadcasting Corporation (RBC), which controlled the nation's radio and television stations. To show the public the degree to which they were being censored, newspapers started publishing large white spaces where text had been censored and removed. Smith soon retaliated with a policy known as "D-Notices." These were originally used by Britain during World War II to make it illegal to indicate that censorship had taken place. Smith's government was now implementing this diabolical policy on the pretext of fighting communism. The blank spaces in Rhodesia's newspapers suddenly disappeared, giving no indication of censored content. People soon forgot that party political censorship was occurring.

Furthermore, to decrease people's accessibility to the British Broadcasting Corporation (BBC), Smith had masses of low-cost, FM "Supersonic Radios" produced that could only pick up local stations, giving the RF near-complete control of information being transmitted to Blacks. Soon, most Whites were using the cheap radios too. Perspectives that differed from Smith's party were mainly disseminated via word of mouth and a few uncensored but obscure news reports.

In 1966, I gained a good friend, Pat Bashford, one of my clients in my ecological consulting business. Pat was a tobacco farmer transitioning to cattle. Like so many others in Rhodesia, he was diversifying to survive the economic sanctions imposed on us by much of the world. Pat also led the multiracial Centre Party, which opposed the politics of Smith and the RF.

During the day, I worked with Pat on his farm. In the evening, over many a whisky, we would discuss the political situation. Pat urged me to enter the political arena and join the Centre Party, but I was extremely reluctant. Not only was I still trying to remain apolitical, but I could see that Pat and others in his party, including former Commanding Officer Putterill (the general who had called out then Colonel Walls for distributing Christmas hats reading "RLI for UDI") were already completely neutralized by Smith's cunning actions. Through his control of the media, Smith had successfully branded the Centre Party and those within it as liberal supporters of Black communist rule—a mantra Smith repeated to the end of his life.

Pat and I despaired over how easily those of us who believed in democracy were branded as liberal idealists supporting the spread of communism throughout Africa. Smith wrote in his memoirs, "One could not have any respect for people who had jettisoned democracy and all the basic freedoms and standards of decency that the Commonwealth once stood for and fought for."[7] Smith swore "his Rhodesia" would be the last bastion of Western civilization in all of Africa. These were fierce words, but Smith was in fact behaving toward all Rhodesians, Black and White, in precisely the manner he so denounced in the "communists." Yet Smith's supporters cheered him on.

From every angle it was a messy situation. The White population that essentially ran Rhodesia and its finances independent of Britain was still very small—at its peak about 243,000 men, women and children following an influx of immigrants after World War II. The Black population was just under six million, comprised of close to 80 percent Shona and approximately 20 percent Ndebele. Other lesser tribes made up the difference. In total, about fourteen languages were spoken in Rhodesia. Our country also had families of mixed race and tribe and a small population of Asians (Indian and Pakistani). Developing a modern state out of this kaleidoscopic demographic was clearly not going to happen under Smith's racist agenda.

---

[7] *The Great Betrayal: The Memoirs of Africa's Most Controversial Leader,* by Ian Douglas Smith (London: Blake Publishing Ltd., 1997), p. 149.

CHAPTER 38

POLITICAL ALLERGIES OR AMBITIONS?

Becoming a Member of Parliament (MP)

1968–1974

I doubt anyone had ever entered Parliament more ignorant than I was about parliamentary procedures.

The "Under 40s" Rhodesian Front members of Parliament following the 1970 election. I'm seated in the center. *(Photo provided from my personal archive.)*

In the late 1950s, by a campfire in Northern Rhodesia, renowned British ecologist Sir Frank Fraser Darling (or FD) had listened to the visionary intentions of the game department's youngest recruit and predicted that I would end up doing research as an academic. Or I would go into politics—a prophecy I rejected at the time.

However, in both the Northern and Southern Rhodesian Game departments, I had come to see that neither the goals of the institutions, nor of most of the individuals within them, embraced the need to determine the causes of, and find a solution for, the widespread degradation of wildlife (and human) habitat. Hunting poachers and shooting "problem" animals, day after day, was not going to save Africa's wildlife or environment. Changing institutions—and the hearts and minds within them—was pure politics.

Discussions with Fulbright Scholars Archie Mossman and Ray Dasmann—as well as other scientific colleagues—often concluded that all might go well if only we could get an ecologist into Parliament to influence policies. In fact, only a few years after FD's pronouncement, I had had to accept that my best ideas—and those of others dedicated to saving wildlife—foundered when we presented them to senior civil servants. Rather, government officials trimmed their sails to the political breezes and furthered careers at the expense of our mission and purpose.

At the same time, my path as a soldier also was leading nowhere that I could call success. My obsession with all facets of irregular warfare had led me to carry on as an apolitical cog in the heart of a machine. Personal ambitions, careers, and politics dominated the military too. I realized I would have to start rethinking at least my allergy to politics.

But all of this was not yet enough to tip me in the direction of political pursuits. What was to do so was my mounting concern about the clumsy strategic handling of Rhodesia's rapidly escalating civil war by both our military commanders and the cabinet of Prime Minister Smith, who was effectively crushing all civic opposition, whether Black or White.

One morning in early 1968, while shaving, I came to terms with the fact that Rhodesia was failing—everything we believed in and were so proud of was about to be lost. I pondered what I would say to my children, Jim and Meg, if in the future they asked me what I'd done in the prime of my life to prevent that loss. I thought about my discussions with Pat Bashford on the idea of penetrating Smith's party and instigating change from within—and how we had dismissed doing so as undemocratic, too difficult, and certainly too dangerous. When I turned away from the mirror that morning, I drove straight to the RF's headquarters in Bulawayo, asked for registration forms from a woman who had no idea who I was, and joined the party.

I informed my father of my motive, but no one else (not even Pat, who was left to guess what I was up to). While I had never in my life heard my father speak ill of any man, when I confided in him why I had joined Prime Minister Smith's party he warned me, "Be careful, Allan—Smith is a pathological liar and extremely vindictive." When I asked why he felt this way, he recalled his time as Secretary of Water Development, during which he observed Smith lying in a water court case while under oath. When the judge ruled against Smith, Smith simply ousted the judge. I was to experience this same level of vindictiveness, but I didn't know it yet—and so I forged ahead.

By the time those at RF's headquarters in Salisbury realized who their newest recruit was, it was too late. On my next call with my clients Colin and Jenny Bickle, Jenny cautioned me that her father, George Rudland (minister of agriculture), had demanded to know who'd let "that little bastard" into the party. I told her to tell her father to relax—I had concluded that they were right, and I wanted to work with them.

Coincidentally, within days of joining the RF, the member of Parliament (MP) for Matobo constituency in southern Matabeleland (an area which included my childhood school, Plumtree, and several of my ranching clients) resigned. An unexpected by-election (special election) was to be held to fill the vacancy. No one opposed the RF, and it was a foregone conclusion that whomever the party nominated would take the seat. The party announced their choice: the Chairman of the Matabeleland branch, John Strong—who was also endorsed by Smith. In the background, I quietly went around to friends and achieved the required number of signatures for a nomination. Suddenly the RF committee had to decide which candidate they supported—their nomination or me, virtually self-nominated and with no support within the party!

Petrol being rationed, Chairman Strong, along with myself and the Chairman of the Matobo branch of the RF, drove in one car to Kezi to meet with the RF committee and the constituency's members, largely farmers. It was a long, silent drive; we arrived after dusk at the Kezi Farmers Hall where about forty people awaited us. Rather than go into the hall together, I hung back, politely offering Chairman Strong the chance to go first. He thanked me and went in with the other party officials. I stood by the door listening, alone in the dark. Committee members asked for all manner of assurances from Chairman Strong about what he would do for their roads, schools, farms, security, dams, irrigation, communication, and more. Strong responded with many promises, as politicians tend to do. Then the committee called me in for questioning.

Their questions followed the same pattern, but my responses did not. I professed I was not a politician and did not like politics. I had no faith in politicians and their endless never-carried-out promises. I reminded them of my years of publicly fighting the government over land degradation. When asked, "What will you do for us?" I said I would make only one promise: "As your MP, I will not be a rubber stamp for the party; I will truly represent *you* and fight like hell for you. If you don't like what I do, oust me in the general election in the coming year." With that, I was voted MP of the Matobo constituency in southern Matabeleland—winning by four votes. The drive home that night was very strained indeed.

Despite my convictions in front of the committee, I felt clueless—and frankly terrified. At age thirty-three, I was the youngest MP by ten years, and I doubt anyone had ever entered Parliament more ignorant than I was about parliamentary procedures.

At this time the 1961 Constitution was in effect, which was intended to gradually increase Black representation in Parliament. Under it, the MPs were divided into two rolls, the "A" roll and the "B" roll. There were fifty A-roll constituencies and sixteen larger B-roll districts, with a complicated mechanism of "cross-voting" that allowed B-roll voters to slightly influence A-roll elections and vice versa. Those in the B roll did not have to meet as many qualifications (based on education, earnings, and property), the intention being to make it easier for prospective voters to enter the political system. Though some Whites only qualified for the B roll and some Blacks qualified for the A roll, in practice the A-roll-elected members forming the government were all Whites and the B-roll-elected members were all Blacks but for one White, Dr. Ahrn Palley. My seat was one of the fifty A-roll seats, all of which were held by White RF members.

The RF MPs were also part of various caucus committees that reported to the main caucus. I joined every committee I could—among them defense, agriculture, and finance—to learn as much as possible about the government's thinking. While Parliament was in session, the main caucus met every Wednesday morning. These meetings, chaired by Prime Minister Smith, were officially secret—no notetaking nor minutes were allowed. If a dispute arose over something agreed upon in a previous meeting, whatever the majority recalled as "right" carried the day. I laid low by primarily sticking to agricultural or conservation issues, as I dared not express my political views. Dr. Ahrn Palley, an elderly and wily White politician opposed to Smith's racism and holding a B-roll seat, soon recognized my inexperience and noted my opposing views—at least those on agriculture and wildlife management. He became a much-needed adviser to me on the ways of Parliament.

In addition to being a novice on parliamentary procedures, I was terrified of public speaking. While I had talked in public on matters about which I was passionate, I had no experience in debate nor giving speeches—as army officers we had mostly been taught to give orders. With great trepidation and shaking hands (which fortunately the parliamentary record does not reveal), I finally gave my obligatory opening speech on a politically neutral issue concerning natural resources.

In the same year I joined Parliament, my adversary George Rudland stepped down and David Smith, a wealthy tobacco farmer, became the minister of agriculture. Phillip van Heerden was still minister of land and water development, and though we had clashed in my sugar farming days, I regarded him a decent man (especially once he started to have to tolerate my apolitical speeches). Minister van Heerden's portfolio included national parks, wildlife management, and conservation—the only issues, other than agriculture, that I felt safe to talk about and was comfortable publicly addressing.

I believed the Department of National Parks and the (much bigger) game department should be one department headed by a professional ecologist or biologist. Though the senior civil servants within the national parks and game department opposed my view, Rhodesia as a whole was apathetic. I committed to changing this and pursued a one-man strategy to generate "a lot of public opinion" that would help bring it about. I started by writing a series of short letters describing what was happening to wildlife and why "we, the public" demanded a parliamentary inquiry (although there was no such demand). I sent these letters to friends and clients around the country whom I knew cared about wildlife. I urged each of them to modify the wording to their style but not the message, then post the letter under their name to their nearest newspaper (in Salisbury, Bulawayo, Umtali, and Gwelo). I gave each a schedule so all the letters would not appear at once. In Parliament we received the various papers in our lounge daily, and I collected the "many" letters of concern from the public "all around the country," sharing them with other MPs as evidence of public concern and emphasizing the need for an independent commission of enquiry.

That worked, and when I had rallied enough MPs, I raised my request for the commission in caucus, and poor van Heerden had to accept at least an enquiry, resulting in the Petrides Commission of Enquiry into Wildlife.[1] This commission, as was inevitable, recommended merging the departments. I then recommended my former colleague Graham Child (an ecologist who was the wildlife research officer in my Southern Rhodesian Game Department days and now had his PhD) to lead the single, united Department of National Parks and Wildlife Management, and it subsequently moved under his professional management.[2]

I gradually gained a feel for how Prime Minister Smith, his cabinet, caucus, and party were running the country, however it certainly came with its challenges. I had to say something on the Parliament floor to show I was not asleep, yet there was only so much I could say on conservation issues, and I certainly would not speak in favor of any of the RF's increasingly racist policies.

I avoided appearing in public situations where it was likely I'd be questioned about RF stances, and I was inwardly ashamed to be part of a party that was destroying so much of what I loved about my country. Although we were regularly urged to recruit new RF party members, I never recruited one.

Regarding my constituents, I represented them as best I could, so much so that constituents of other MPs began approaching me to help them, confiding that their own MPs were, in their view, simply party hacks. This was truly welcome, because I had to survive an upcoming general election in my Rhodesian Front guise.

In 1969, a new constitution was proposed by the RF and of course approved in a referendum, there being no other possibility. Gone were the "non-racial" A and B rolls; the changes essentially prohibited Black Rhodesians from ever voting for or against the

---

[1] The Petrides Commission of Enquiry into Wildlife was named after George Petrides, a professor of wildlife management at Michigan State University in the United States. I met Petrides at a conference in Salisbury where I presented a paper. Afterward, Petrides approached me, saying my presentation was "the most advanced wildlife management thinking he had heard of" and asking if he could work with me and the game department. I explained that I was in fact in the process of being forced out of the department because of my thinking. Petrides and I remained in touch and when the need arose for an international person of repute to head the commission of enquiry, I put his name forward to Prime Minister Smith.

[2] Though merging Rhodesia's game department with the parks department was one of my few small victories in Parliament, it did not lead to a significant improvement in the country's wildlife management policies. The lack of professional managers wasn't the problem; it was our lack of knowledge about the underlying cause of the habitat degradation—of which I still had much to learn.

government. The new constitution reserved fifty parliamentary seats exclusively for Whites and only sixteen "side bench" seats[3] for Blacks. Rhodesia's population of about 230,000 Whites, plus another 8,000 Asians (Indian and Pakistani), and 15,000 Colored (mixed race) people, could vote for the White seats (guaranteeing a White government). However, Blacks, of which there were just under six million, could only vote for eight of the sixteen side bench seats reserved for Blacks. The White government would appoint the other eight from amongst the tribal chiefs who were unquestionably under government control.

In the election of 1970, under the new constitution, the RF won all fifty seats reserved for Whites, just as Pat Bashford and I thought might happen with Prime Minister Smith's control of the military, police, judges, and media. Black Rhodesians of course filled the eight side bench seats they were allowed to vote for, and the remaining eight seats reserved for Blacks were filled by tribal chiefs. These chiefs were chosen by the Ministry of Internal Affairs (formerly Native Department), now headed by Hostes Nicolle,[4] an extreme racist in the mold of Smith.

The sixteen Black members could never form a government—a majority voting block—when fifty seats were reserved for Whites. Even White political parties opposed to Smith had no hope of ever gaining a single seat. With all side bench seats now only for Blacks, no White MP could be elected to one, as my mentor Dr. Ahrn Palley had been, so he was out. My friend Pat Bashford ran but was beaten by a large majority, as were all Centre Party members but for those Centre Party members who were Black and hence running for the side bench seats. I, as an unopposed RF candidate, automatically retained my seat and was safe for at least a little longer.

Where there had been a ten-year age gap between other MPs and me, we now had an influx of younger MPs. The new men (there were no women) were of my generation, but not—I believed—truly representative of it. They had been selected by the RF party as men who believed, like Smith, that Whites should and would control the government of Rhodesia "for a thousand years." I had tried "sounding out" some of these new MPs, saying I was grappling with some of the logic—*why did we not worry about whether our lips touched a glass handled by Black servants, yet we went to great pains to have separate toilets so that our backsides wouldn't touch the same toilet seat?* And I'd never heard of a White soldier—when undergoing a blood transfusion—ask whether he was getting Black or White blood. *Were they too struggling to reconcile these observations?* I never did get sensible answers.

Sadly, the discourse during parliamentary debates worsened, frequently succumbing

[3] Side bench seats were a new concept and unique to the Rhodesian Parliament. Having them enabled the Smith government to say that Blacks "had the vote," but the sixteen side benchers were basically disempowered.

[4] William Hostes Herault Nicolle was described by Ken Flower in his memoir—*Serving Secretly: Rhodesia's CIO Chief on Record* (Alberton, South Africa: Galago Publishing [Pty] Ltd., 1987, p. 40 and pp. 87–88)—as "an advocate of apartheid" who "turned the Department of African Affairs [the Native Department] into a political weapon for the government's use."

to verbal brawling and racial insults. RF MPs would interrupt side bench MPs with an onslaught of racial remarks and ridicule, inviting retaliation. So embarrassing and unconstructive was this behavior that I wrote a private note to Prime Minister Smith, drawing his attention to what I thought was doing great harm and informing him that almost all the abuse was being initiated by our own RF MPs. Might he have time to sit in chamber to notice this behavior and end it? Smith did not respond to me, however, at the next caucus meeting he repeated almost verbatim the content of my note and requested a stop to the abusive interactions.

For a while the atmosphere on the Parliament floor improved; there was more listening, less shouting, and even some humor. Then one afternoon when a Black MP was speaking, Smith interjected with a string of racially offensive comments, inciting other RF MPs to join him. He was behaving in the very way he had two weeks before appealed to his RF members not to behave. I walked out in disgust.

Although I strongly opposed Prime Minister Smith's views and believed he was betraying all moderate, non-racist Rhodesians, Black and White, I could never hate the man. Smith was charismatic (and there was a side to him that I appreciated; for example, even though his son, Alec, sided with me and opposed his father's views, Smith never let this affect his love for his son). Many of Smith's opponents claimed he was corrupt and supported this with stories that he and some of his ministers were siphoning money out of the country. Later (when I was in a role to publicly oppose Smith's government—see Chapter 39) people pleaded with me to attack him as corrupt and I refused as there was a lack of evidence. To the end I've never believed Smith to be corrupt (unlike many of his ministers and supporters, he remained in the country after the war ended). Rather, I saw him as a diehard racist, and a short-sighted, gut-level politician—able to lie with an incredibly straight face—who confused the Cold War threat of communism with the demise of colonialism. The only way to combat this communist threat—in Smith's view—was through ruthless control of a democratic government.

By the end of 1971, our civil war was predictably escalating, and I was spending more time on operations with the army. I had recently led one of the TCU tracker teams into a successful "contact"[5] after a series of setbacks, and Smith had heard about it. At a subsequent caucus meeting, Smith went to the unusual step of singling me out for praise as an RF MP engaged in combat. Soon after this, a couple of MPs passed on a rumor that he might include me in his next cabinet, and I was horrified. I could not carry my deception that far.

The time had come to act. I was no longer convinced I could achieve change by working with Smith from within the party. Smith had always said—and repeated often in his memoir, *The Great Betrayal*—that as Rhodesians we must sort out our own problems.

---

[5] *Contact* was the Rhodesians' term for the type of skirmish typical of our bush war—a brief battle at close quarters in dense brush and tall grass that would ensue after successfully tracking our opponents.

I also believed that we had to sort out our problems, but Smith had no intention of negotiating with our Black political party leaders, ever. And that meant many more lives would be lost before he was forced to do so from a weak position of total defeat.

Every success of our security forces only led to Smith becoming more emboldened and intransigent. The RF propaganda machine made things worse, labeling Mugabe and Nkomo, the most effective African political leaders, as "terrorists"—which riled far more reactive emotion than "guerrillas." On top of this, party censors worked to keep their White constituents hot for war, intentionally underreporting the army's successes and instead magnifying the atrocities committed by the "terrorists." For example, on the same day that the army would wipe out a group of "terrorists," somewhere else the "terrorists" might have attacked a soft target (all they *could* do) and slaughtered everyone in, say, a mission station. However, the army success would be censored and only the "terrorist" atrocity would be broadcast, unintentionally spreading terror and undermining investor confidence, but keeping Whites loyal to Smith.

Against the backdrop of this propaganda, Smith would then boast that his government was winning the war because of the number of terrorists killed by Rhodesia's security forces each month. But those numbers were meaningless when government was at the same time losing the support of millions of Blacks in the rural areas. The government offered nothing but restrictive, racist laws—a far cry from the protection Blacks needed whenever ZIPRA and ZANLA usurped the power of their chiefs (many of whom did not support the guerrilla forces[6]) in brutal nighttime attacks. Later, when free to speak, I often said that if I had designed a media control policy to ensure self-defeat and the rapid rise of Black rule, I could have done no better than the one Smith's cabinet and generals created.

[6] Many of the old chiefs considered the guerrilla leaders political upstarts and would not support them. Any of the chiefs' people suspected of being "sellouts" to either guerrilla army—ZIPRA or ZANLA—or to the Rhodesian Army for that matter, were tortured and killed, generally at night after the Rhodesian Army had cleared an area by day and then left, not having the manpower to post a protecting force in every rural community.

CHAPTER 39

THE RHODESIA PARTY

Questioning Those in Power

1972–1974

After five years of observing the inner workings of Smith's government, I could predict their almost every move.

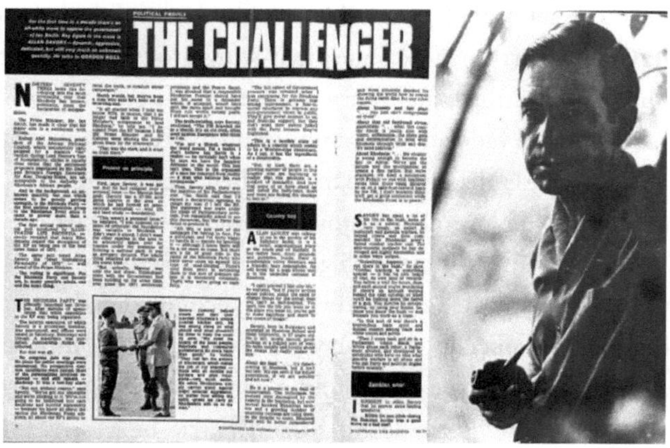

A political profile published in *Illustrated Life Rhodesia* (February 1973).[1] The article quotes an opinion poll that the emergence of the Rhodesia Party (RP) was one of the best news items of 1972 and that I was voted that year's "most stimulating personality." This only angered the Rhodesian Front (RF) further.

In February of 1972, at the next caucus meeting and before any discussion began, I asked if I could make a statement, to which Prime Minister Smith agreed. I then shocked everyone by stating that I had no confidence in Smith nor his cabinet to lead the country out of its dire situation. Hence, I was resigning from the RF party.

Heated discussion followed, first with caucus members trying to persuade me to stick it out and, when I refused, changing their tactics and demanding I vacate my seat in Parliament. To that I replied that I was only resigning from the party and not from my seat. I would be "crossing the floor" despite there being no other party in Parliament that I could cross the floor to join. Amid a slew of angry comments, I walked out of the

[1] *Illustrated Life Rhodesia* was published between 1968–1978. It was banned and ceased publication in 1978.

meeting, went home, and after unwinding from the intensity of the day, immediately got to work. I intended to form a new party, though I had no idea how to start a political party against the most ruthless government ever in Rhodesia. I had no supporters, no one to discuss it with, and no money.

I began by letting Pat Bashford know I had resigned. Anxious to keep the Centre Party functioning, he encouraged me to join them. Though I would never oppose them, I felt I could not join a defeated party. Our now entirely racial constitution, Smith's destruction of any multiracial effort, and his branding of White opposition as either "liberal communists" or "right-wing extremists" necessitated a new kind of party. But what sort of party would Smith have difficulty branding as liberal, communist, or extreme?

I wrote letters to a handful of people I knew who had political clout but were not members of the Centre Party nor extreme right-wing groups. I asked them to invite others of like mind to a meeting in Salisbury. We were to discuss the possible formation of a new party that would evade all labels and simply be Rhodesian, with an intent to procure good governance for all people, no matter their race or tribe.

On a memorable day in November, a group of approximately sixty people assembled in Salisbury. Within a few hours—and with great hope and enthusiasm—we formed the Rhodesia Party.[2] Having convened this gathering, I recommended we not elect officers until we were functioning officially as a party but name a troika (three people working together) to lead the party until we could hold its first congress and elect office bearers. To this everyone agreed. The group decided that the trio be Morris Hirsch (a former MP), Roy Ashburner (who had stood for Parliament as an independent and lost to Smith), and me. Shortly after our first meeting, George Kileff, a tobacco farmer opposed to Smith's views, generously donated $35,000[3] to get the party launched.

We printed our first brochures, enthusiastically spreading the word and began enlisting members. Unfortunately, it wasn't long before self-interests started to surface. Morris Hirsch began jockeying for support to be party leader. In addition, rumors swiftly circulated that many businessmen would not support the party if I became the leader. They favored Roy Ashburner, an older, supposedly more stable man. I tried to maintain the peace (without lobbying and manipulation) until several Rhodesia Party branches had formed around the country and could send representatives to a congress in Salisbury. At this first congress, I was elected leader of the party; Morris and Roy left the party rather than serve under me.

One of my first steps was to arrange that any Black Rhodesians wanting to join the party be directed to me. A string of them approached us to join; however, as planned they

---

[2] We chose the party name, Rhodesia Party, for its neutrality; it wasn't going to provoke anyone, though it had been used some years before by a short-lived and quickly forgotten party.

[3] At the time, the Rhodesian dollar was on par with the US dollar. In 2025 US dollars, the Rhodesian dollar amount translates to $269,150.

first talked to me. I explained how much we valued their support and understanding, then pointed out that if they joined us, they would effectively destroy any chance we had of fighting the RF. Like the (multiracial) Centre Party, Smith would then brand us multiracial, and thus as liberals who supported "Black communists." We would be far more effective—and could avoid ensnarement in Smith's propaganda—if we remained a conservative, non-racist White party opposing the racist all-White RF. The interested Blacks understood, and word got around; Black Rhodesians—both inside and outside of Parliament—did not try to join us but gave us help in every way they could.

Though I accepted my election as president and thus party leader, I would have preferred to remain behind the scenes, especially after the party's first public meeting at Bulawayo City Hall. The largest hall in the city, it was packed with a standing-room-only crowd and those who couldn't get in were crammed around the doors outside, straining to hear the proceedings. I received more than one prolonged standing ovation, however the effect on me was horror, not pleasure. I suddenly truly understood the power mass adulation has on the human psyche, and I did not like it. Later—on a high that seemed unhealthy—I battled to sleep. I now knew what Kipling meant in his poem "If" when he wrote, "If you can talk with crowds and keep your virtue . . . "[4] I better understood what happened in Nazi Germany when politicians grew an unsatiable hunger for the power served up from the crowds—how dangerous that is and how easily it cripples our humanity. Alun Vaughan—who handled Rhodesia Party publicity and was one of our sharpest and most politically astute minds—had a habit of popping into my office at the end of each day to chat. One evening he mused, "Allan, I have finally figured you out. You genuinely don't want power."

Despite its overwhelming attendance, little was publicized about the largest political meeting since the Unilateral Declaration of Independence (UDI). All that was reported in the papers and in broadcasts the next day was a statement saying that at a political meeting in Bulawayo I had been asked a question by the audience. The answer that was printed in the papers was the opposite of what I'd given. (The questioner was in fact an RF plant who I had called out, saying—before answering his question—that the question was not his but was on the piece of paper in his hands, given him at RF headquarters.) Nonetheless, word of the Rhodesia Party grew, garnering a significant support base in the Bulawayo area.

In 1974, the main African nationalist leaders Nkomo, Mugabe, and Sithole were released after a decade of detention as political agitators. The Rhodesia Party hoped they would remain in Rhodesia, operating politically. However, fearing re-detention, they immediately exited the country to join their parties from secure bases in neighboring Zambia and Mozambique.

[4] This poem was first published in *Rewards and Fairies* (London: Macmillan and Co., Limited, 1910).

Aside from the government, the two main political parties in our civil war—the Nkomo-led ZAPU and the Mugabe-led ZANU—operated two separate guerrilla armies, ZIPRA (Zimbabwe People's Revolutionary Army) and ZANLA (Zimbabwe African National Liberation Army), respectively. Predictably, Russia and China were arming and financing these armies because no Western power would do so. Logically they started calling one another "comrade," quoting Lenin and Mao Zedong—further convincing Smith that he was fighting communists.[5]

Based on pure intuition, as I had no evidence, I felt Mugabe would be the most likely Black leader to emerge victorious. While in detention, Mugabe had completed two law degrees and executed a coup against Sithole, the detained leader of ZANU. He thus became head of the party and the corresponding military wing, ZANLA. I sought out Mugabe's younger brother, Albert, who was still in Salisbury, and befriended him in the hope that he could connect me with Mugabe. This connection was short-lived; not long after I got acquainted with Albert, he was mysteriously found dead in his swimming pool.

I also sought out Josiah Chinamano,[6] Nkomo's second-in-command, who on release from detention had been confined to within a certain radius of his home in Salisbury. I invited Chinamano to dinner, and we proceeded to discuss many an issue. At one point I asked Chinamano why he was the only prominent Shona supporting Nkomo. He asserted that Nkomo was the most promising future leader, and government should not be tribal. We became friends and allies, bonded by our common concerns and fears. This new friendship was a bridge to connecting with others.

After five years of observing the inner workings of Smith's government, I could predict their almost every move. Inevitably, in refusing to leave Parliament but instead forming another party, I was swiftly condemned by Smith. He personally visited Bulawayo, addressing the Matabeleland branch of the RF with such bitter remarks about me that several men walked out. Those men and others began informing me about their party's moves against me. Predictably, the RF incited a nationwide smear campaign. The doctoring of my words in newspaper reports and personal attacks over television and radio went on for years, even after I'd left the country. All that Smith postulated "communists" were doing to silence dissent he now did tenfold to me.

The emergence of the RP was perhaps the greatest threat faced by the RF since coming to power, on the surface because it was an all-White party that was gaining in popularity with White voters and difficult to brand liberal or multiracial. The party was also speaking about a way forward that was beginning to make sense to a war-weary

---

[5] Vladimir Lenin, as leader of the Soviet Union from 1917 until his death in 1924, created a one-party state under the Communist Party. Mao Zedong founded the People's Republic of China and served as the chairman of the Chinese Communist Party from 1943 until his death in 1976.

[6] Josiah Mushore Chinamano was also detained by the Smith administration in 1964 and spent a decade in detention. He was second-in-command to Joshua Nkomo in the ZAPU party.

populace. This infuriated the RF, Smith in particular, and reporters and historians took note.

In his *History of Rhodesia*,[7] Robert Blake wrote:

> *It is perhaps this intense "Rhodesian patriotism" which makes [Smith] almost unnaturally sensitive to criticism, although, to be fair, such sensitivity seems to be a characteristic of most Rhodesian Front politicians. I can remember the surprise with which in 1973 I heard him later at a dinner party denouncing at great length in extraordinarily bitter language an RF member of the Assembly, Alan [sic] Savory, who had just renounced the party whip and proclaimed himself to be independent. There was something obsessional about the way he spoke. Savory's personal "UDI" was obviously not appreciated, yet the RFs strength in a House of 66 was only reduced from 50 to 49, and it seems odd that a Prime Minister, who previously held all the European [white] seats, should have been disturbed to find himself instead holding all but one.*

[7] From Robert Blake, *History of Rhodesia* (New York: Alfred A. Knopf, 1978, p. 362).

CHAPTER 40

WHEN "WAR" GETS PERSONAL

The Politics of Love, Loss, and Land

1968–1974

The government's attacks on me intensified on all fronts—military, political, civic, and scientific.

Our home in 1971, a year before Kazuma was expropriated. Shirley and Meg are having tea in the garden while Rodger toddles across the lawn.

Throughout my years in Parliament, I continued to juggle my several lives, from my service in the army, to my consulting business to my independent research. It was a difficult and stressful time. My service as MP added even more travel to my already hectic life; thankfully, my personal plane helped tremendously. Having firsthand experience with the violence erupting in my country (and the fact that I was not only using my plane to fly to and from Parliament but to fly to and from forward airfields[1] when on army operations) I took additional caution. I was almost always armed with a hidden weapon, and to avoid becoming a target in the air for heat-seeking Strela missiles, I had my plane painted with the grey "Strela paint" used by our air force to render aircraft less detectable.

Try as I might, I never did achieve a good balance. In the early years I might be one day on Kazuma with my family and then get notice from the army that there had been a guerrilla incursion. I would have to whisk Shirley and the children to a place of safety

[1] Forward airfields were bush airstrips as close to areas of combat as practical, used for refueling and re-arming support aircraft.

and immediately fly off to the nearest forward airfield. There, I would meet up with my TCU officers and men, who had also had to mobilize with short notice. After a few days of tracking guerrillas in the bush, I might then head to Liebig's Ranch to discuss game ranching matters with one of my TCU corporals, several of whom also worked for me and oversaw our WUS hunting camps. From Liebig's I might then fly to Salisbury to attend a parliamentary debate for a couple of days. Then I would fly off to Swaziland, South Africa, Botswana, or Namibia to advise my rancher clients on their management.

Each week I would shuffle between these widely varied commitments, never seeing my work at one place as distinctly different from the work at others. It was all related to the refrain that played constantly in my head: *poor land leads to poor people, social breakdown, violence, and war.*

As is apparent in this memoir, I was obsessed with serving my country and addressing the cascading issues we faced. This obsession was sadly at the expense of my own family; I believed so deeply that unless everyone's family was secure, no one's family was secure. Shirley and I started to drift apart, even more so after our son Hugh's death in 1968. Two years after that tragedy, we had our third son, Rodger, and in 1973, our second daughter, Claire. By the time Claire came into the world, however, our marriage was over.

We divorced, and Shirley left for South Africa with Jim, Meg, Rodger, and Claire. I didn't attempt to gain custody as I wanted to avoid any bitterness and blame (not to mention under Roman Dutch law, as we were, fathers rarely got custody unless the mother was clearly irresponsible and unfit to raise the children). What I had not foreseen was that my children would move so far away, and the distance—in addition to a mother's bitterness—would create lasting effects.

I soon crossed paths with a woman named Janet Barnes when she volunteered as the new party's secretary. She was professional and proficient, having worked as Secretary of Defense Barney Benoy's private secretary. Her family published the *Gwelo Times* newspaper,[2] a key source of news that served the town of Gwelo, located in the midlands of Rhodesia. Janet and I courted to a backdrop of political turmoil and were married in 1975.

It was only after I got to know Janet that I learned about secret plans at the highest levels of our military forces, which included the police (BSAP), army, and air force. Janet shared a harrowing story of her personal experience with the dissonance at the time of UDI in 1965. The officer commanding the BSAP at the time, Basil Spurling, supported UDI and Prime Minister Smith's cabinet, but the commanders of the army and air force were in favor of standing by their oath of allegiance to the 1961 Constitution, not to a political party. These two commanders, together with high-ranking civil servants loyal to

[2] The *Gwelo Times* was edited by Janet's mother, Margaret Barnes, and was one of the last papers to hold out against Prime Minister Smith and his secret censorship.

country and constitution, were colluding on possible action plans, the details of which were secured in Benoy's safe in Salisbury. The plans included a possible role by Evan Campbell, Rhodesia's last High Commissioner to London, and the potential detention of Smith.

Smith caught wind of the collusion, and when Benoy and Campbell were in London, he conveniently made a visit to Benoy's office. Accompanied by Clifford Dupont (the RF-appointed president who replaced British-appointed Governor Sir Humphrey Gibbs), Smith entered Benoy's office and demanded that Janet open the safe. Janet, knowing the safe's sensitive contents, refused. This angered Smith and Dupont, who thundered, "Do you know who we are? We are *ordering* you to open the safe!" After continued intimidation, Janet acquiesced, but only, she said, if Benoy gave his permission. Smith and Dupont agreed to let her talk to Benoy in their presence.

Janet promptly phoned Benoy in London, however Campbell answered instead. Campbell, astutely realizing that something was wrong, said Benoy (who was standing beside him) had already left for Heathrow. Smith and Dupont departed in frustration, but not before locking up Benoy's office and declaring they'd soon be returning to forcibly open the safe.

In near panic Janet rushed to her senior colleague Arthur Hasler and explained the situation. She gave Arthur the combination to the safe, with which he quickly climbed out of her window and into the window of Benoy's locked office, opened the safe, and scrambled back through Janet's window with the incriminating documents in hand. Janet then took the documents, drove out to the Highlands suburb, and buried the papers in her sister's garden. On her return, she found that someone had indeed blown the safe open. She then phoned Benoy's wife, advising her to get to the airport and apprise her husband of the situation before he spoke to anyone on his arrival.[3]

The government's attacks on me intensified on all fronts—military, political, civic, and scientific. At times, when I had been out of the country, I would return home to find Janet in tears after being bullied by senior army and air force officers at garden parties. They would sidle up to her, asking, "What's it like being married to a traitor?" Interestingly, however, when I was present, the same people who had tormented Janet often approached me, quietly asking, "Allan, do you really think we're losing the war? What would you suggest we do? Should we be getting our children out of the country?"

Emboldened no doubt by my unpopularity, the Department of National Parks and Wildlife Management embarked on a mission to block private wildlife research and shut down WUS, our game ranching and research company. The department sent a new biologist, Tony Farrar, to report on our operations and to do a survey of the eland population

[3] The incident with Benoy's safe was confirmed by Ken Flower, head of Rhodesian intelligence (CIO), in his memoir, *Serving Secretly: Rhodesia's CIO Chief on Record* (Republic of South Africa: Galago Publishing [Pty] Ltd., 1987), p. 42.

in the large Gwaai Forest that WUS leased from the Forestry Commission. This came at a time when our game ranching efforts were showing impressive results. Eland and sable were increasing in both numbers and productivity, most females now surviving to breeding age whereas originally approximately 80 percent of them had died before reproducing.

Farrar, upon arrival, dismissed our technique (developed by Archie Mossman) for assessing population health and productivity (by age structure rather than *numbers* of animals sighted) as "indigestible." Instead, he chose to drive around the Gwaai Forest (nearly half a million acres) and count the elusive eland from his Land Rover.[4] His count? Nine eland. Based on this, Farrar claimed we were over-shooting and cancelled the permits for WUS.

WUS was forced to file for bankruptcy. Robbie Robertson, who was managing the company, oversaw the filing but first asked Archie Mossman and me to resign. It wasn't until after Robbie shockingly committed suicide that we realized he had prompted our resignations to clear us of any financial ties to WUS. He had taken full responsibility for the bankruptcy, which was not considered a light matter in Rhodesian society.

The Department of National Parks and Wildlife Management was not finished with its attacks; they also moved to have the government expropriate my land at Kazuma. I had offered the department two seats on my board and notified them that, according to my will, Kazuma would become a national park upon my death (Kazuma was ideally located near the Victoria Falls, a major tourist attraction, and the only property in Rhodesia having animals as diverse as gemsbok[5] and oribi[6]), but none of this mattered.

Friends within the department told me the reason being given to expropriate Kazuma was to create an immediate national park of it, disguising the fact that they simply wanted to shut down private wildlife research (and consequently me). When I made this public, the department quickly pivoted and stated they were expropriating *all* private deeded ranches in the Matetsi area south of the Zambezi to create one large national park (this consisted of 1,258 square miles extending from the Zambezi in the north to the Wankie National Park in the south).

My internal informants again confirmed the truth: the department had no intention of creating a large national park. Rather, they sought to form a grand hunting safari area under their control but knew that proposing a national park was more palatable to the public. It was unprecedented for government to expropriate so many private properties—

[4] In being tasked with surveying the eland population, Tony Farrar's method (counting eland sighted from his Land Rover), ignored two known facts: the eland were being hunted by teams using Land Rovers and subsequently had learned to hide upon hearing the approach of vehicles; and of most importance, more than a decade earlier we had established as much as a 100 percent error in direct counting on thirty-five acres during Operation Noah.

[5] Gemsbok are large, light-colored, greyish antelope, very similar to an oryx. Their habitat is wide open, dry grasslands verging on desert. Both sexes have long, straight horns and run in small herds.

[6] Oribi are small antelope weighing only twenty-six to forty-nine pounds. They are commonly found in open short grassland in pairs or small groups. Only males possess horns.

most of which were already game ranches—just to lease those properties to hunting safari concessionaires under National Parks' control.

What the department hadn't foreseen was that the affected landowners would turn to me for help, elevating my fight from a personal one to that of the community. Phillip van Heerden (minister of land and water development) tried hard to defeat us, even going so far as to commission a "Minister's Committee Report" under my former opponent Archie Fraser, who had earlier informed me that over his dead body would private enterprise succeed in the wildlife field in Rhodesia.[7] Fraser and van Heerden's intent was to discredit the conclusions of the Petrides Commission of Enquiry into Wildlife (which I had championed years earlier to help professionalize the merged national parks and game departments). However, when members of the committee visited the Matetsi area, they asked me for advice in writing the report, and together we drafted the recommendation that private ranches between the Zambezi and Wankie National Park be *encouraged to utilize and manage their wildlife, as they were already doing.*

I knew I could not save everyone's property, but I could certainly put up a good fight. Because expropriating deeded land required parliamentary approval, and I was nominally still an MP, I forced a private members debate that became something of a legend. At the odds of forty-nine to one, I kept the "Matetsi Debate" going for two weeks—from July into August of 1972. Every detail was documented in the parliamentary record, the Hansard—the transcript of the debate spanned 806 columns, with two columns per page.

Rubidge Stumbles, a lawyer and brilliant parliamentarian by reputation, was the RF Speaker of Parliament throughout the debate. He would have gladly prevented any debate at all from happening by assuring that no party member would second my motion to open one. However, the RF members were caught off guard by the Black MPs in the side bench seats. There was nothing in the constitution preventing a side bench MP from seconding a private member's motion. The Black MPs helped me by stepping up to do just that.

The legendary debate ensued; the RF MPs only realized at the last minute—and too late—that I had duped them into voting for my property to at least become a national park, as it remains today. The rest of the expropriated ranches, unfortunately, were leased to safari hunting concessionaires—as the national parks department had planned—and they remain so to this day.

In reflecting on my Matetsi Debate success, Stumbles, despite his opposition, must be respectfully acknowledged. A few years earlier Stumbles had taught me a valuable

---

[7] Archie Fraser's comment was recorded in the Hansard (July 26, 1972, Matetsi motion), which contains the transcripts of all parliamentary debates. The Hansard was named for Thomas Curson Hansard (1776–1833), a London printer and publisher, who was the first official printer to the Parliament at Westminster.

lesson. Following a debate, he called me into his chambers and asked if I thought he was biased as RF Speaker of Parliament because he was so frequently ruling my behavior as "out of order." I said yes, I did feel he was both biased and unfair, especially when I was one against so many. He then leaned in and wisely explained, "If you were to observe closely, Allan, you would see that I only rule you 'out of order' when you attack the *motive* of your opponents, not when you attack their *actions.*" I thanked him, walked out, and committed to becoming a more effective debater.

Following the Matetsi Debate, Stumbles again asked to see me in his chambers. When I entered, to my amazement, Stumbles rose, walked around from behind his desk, shook my hand, and congratulated me, exclaiming, "In all my years I've never witnessed a debate where one man, forbidden by the rules to read from notes, defeated so many!"

The "winners" don't necessarily write the history books, however. Today if one visits Kazuma, or reads the Wikipedia entry on Kazuma Pan National Park, they will learn only that "Kazuma Pan was proclaimed a national park in 1949 but was deproclaimed in 1964 as no development had taken place. It regained its National Park status under the Parks and Wild Life Act of 1975."[8]

Once Kazuma was no longer mine, I had to argue in front of a court of RF judges for fair compensation. Lindsay Cook, my resolute lawyer, fought hard. Since the law established land value based on what a willing buyer would pay a willing seller, I contacted a buyer in the United States to testify (years prior he had been interested in buying Kazuma as it was incredibly valuable wildlife land close to Victoria Falls, but I had turned him down). The millionaire businessman flew over and repeated his offer for Kazuma, right there in front of the court. Not surprising, the RF judges cared nothing about the value of Kazuma. They dismissed the buyer as a "wealthy American" ignorant of how cheaply "cattle ranches" could be bought in Rhodesia. They then decided to compensate me with what amounted to 8 percent of the price the buyer had offered.

Again, the attacks did not stop. The government's legal team now set the date at which payment for my land would legally take place, and the date was six months out. In the court they trotted out witnesses testifying to my military record. Based on the testimony, the officials reasoned that I was not likely to survive another six months of the war and so the value of my land needed to be reduced. The judge duly agreed, reducing Kazuma's value by another \$25,000.[9] Neither the judge nor anyone from government would look at my lawyer and me; their shame was palpable, but the matter was settled.

That night, with friends in the SAS mess, we again drank to "My Fucking Contribution" as the only operational officer who had not only been awarded an MFC (Military

[8] This Wikipedia entry is per my access on October 6, 2025.

[9] At the time (1972), the Rhodesian dollar was on par with the US dollar. In 2025 US dollars, the Rhodesian dollar amount translates to \$192,250.

Forces' Commendation, Non-Operational) but had also been forced to pay so much to serve his country.

More was to follow. To prepare for the Matetsi Debate, I had written to the national parks department requesting a copy of an internal report they had written about Kazuma. I knew the contents, as I had helped in the writing of it, and knew that nothing was secret nor confidential. I also considered it normal to ask for nonclassified information and did so regularly, as did all RF MPs.

In November of 1972, while I was engaged in consulting work on Sandy Firks' farm (located in Umvukwes, about 100 miles north of Salisbury), a police Land Rover rolled up with two men from the Criminal Investigation Department (CID). Assistant Commissioner Ron Eames approached us and announced he was on official business: I was to be charged under the Official Secrets Act for requesting a government report from a civil servant.

Once more, I had to face an RF-appointed judge. I provided all the requested evidence, never denying I had—as a Member of Parliament—asked for a report that I had provided consultation on as it was being written, nor that I had wanted the report for use in a parliamentary debate. I was convicted of "inciting a civil servant to break the law" and fined $500.[10] Thankfully, even that judge could not bring himself to sentence me to jail time. Had he done so, I would have been thrown out of Parliament, which was of course the RF's intent.

The public reaction was noteworthy; few spoke out loud against these repeated attacks on me (I believe most felt ashamed of the RF's behavior but continued to endorse Smith out of fear), however I sensed quiet public support—a support that even extended to South Africa. On one occasion I visited my father, now in his seventies, in Durban, where I dined with him and several of his friends at the top of a tower overlooking the city. When I requested the bill, the waiter said there would be no charge. The owner did not want to be seen talking to me but had recognized me and appreciated all I was doing; the dinner was on him.

And I'll never forget the afternoon I was walking along the street from the court to Parliament when a barber, having seen me coming, emerged from his shop and vigorously shook my hand, urging me to keep going. It was a rare moment of spoken gratitude.

[10] In 2025 US dollars, the Rhodesian dollar amount translates to $3,845.

CHAPTER 41

SHIT SCARED

The End of My Time in Parliament

1973–1974

It took more than a decade of civil war for me to learn that men will commit acts of bravery that are in reality cover-up for lack of moral courage.

Being interviewed at my Salisbury home by the BBC—whose reporters did their homework (unlike the reporting of so many who unintentionally reinforced Prime Minister Smith's propaganda), 1977.

Times were rapidly changing, and Rhodesian Front (RF) supporters were becoming increasingly fearful because of it. Some of the government's actions made so little sense that their supporters began to suspect malicious powers were now involved and conspiracy theories began to circulate—the Rothchilds, or one-world plotters were determined to bring us down, and so on. When asked about this, I said there was no conspiracy, that those people didn't care about us. The problem was that fifty men meeting in secret (as Smith's caucus did), and thus anonymous and unaccountable, could produce some very stupid decisions. A case in point: the decision that precipitated the closure of the Zambian border.

On January 10, 1973, I was asleep when my phone rang just after midnight. It was the BBC. They had the habit of ringing at all hours to get me on their news programs, but this time was different. They asked, "What do you think of Smith closing the border with Zambia in retaliation for the most recent incursion?"

Although only half-awake, I gave my opinion: "Well, that's extremely stupid." Closing our rail and road points of entry from Zambia was an act of bravado that made no sense. Although Rhodesia had recently been penetrated by guerrilla gangs, those gangs were mainly ZANU soldiers from Mozambique not Zambia and never came through the border posts with stamped passports, but rather through the bush where border control did not exist.

In the dark in my pajamas, I reasoned that South Africa's government would force Prime Minister Smith to reopen the border immediately as they needed the rail connection to Zambia through Rhodesia. Smith, I predicted, would then reopen the border but give a face-saving speech, which he did, saying that President of Zambia Kenneth Kaunda should take the border closing as a warning for harboring terrorists. This would leave Kaunda no option but to keep the border closed or be humiliated publicly, and it would be the beginning of the end. It was essential to Rhodesia's survival to keep the flow of goods from Zambia to the coastal ports open; now everything going out of Rhodesia clearly would be from Rhodesia, not Zambia, and blocked by sanctions, as would everything coming in. In the following days, exactly this followed.

The Rhodesia Party (RP), throughout this period, made progress in gaining support, our manifest remaining fairly apolitical and focused on "good governance." In due course we had by-elections, and I experienced the difficulty of leading a party under an autocratic —rather than democratic—government. Every party member wanted to participate in every election held, whereas I advocated that we should boycott elections that failed to meet basic democratic requirements. After all, Prime Minister Smith and his RF followers—given their ruthless control of the media, judges and more, on top of their relentless propaganda instilling fear of "terrorists"—were guaranteed to take all fifty seats. Any election under such conditions would only empower Smith to falsely keep claiming full White support, further reinforced by inaccurate reporting from the international media.

Coverage of Rhodesia's situation by the international press, particularly that of the US press, was generally of a low standard. Frequently, reporters would tell me they had interviewed Smith and learned that *every* Black Rhodesian could vote for members of Parliament! *With more than five million Blacks already having the vote, why was the world sanctioning Rhodesia and supporting communism?*

I always responded to the naïveté of the international press with questions: "Did Smith also tell you that those five million can only vote for eight of sixteen side bench seats? *And* the remaining eight side bench seats are reserved for chiefs selected by Whites in Smith's government? *And* only Whites can vote for the other fifty seats, thus forming the majority block that governs the country?" I would explain that this political sleight of hand effectively disenfranchised Blacks. I felt I was wasting my time with most journalists, so I declined interviews with anyone other than the BBC, the *Christian Science Monitor,* and Agence France Press. Their reporters did their homework and asked intelligent questions.

Convinced that the RP would make far more impact both internally and externally by boycotting farcical elections, I appealed to the RP to announce we were abstaining from voting under Smith's government. To further strengthen our protest, I suggested that we dress in uniform—whether it be our army, police, or air force uniforms—and surround the Rhodesian Broadcasting Corporation's (RBC) facilities, carrying placards stating that if we must give our lives for democracy, we are entitled to a democratic vote. This I believed would have a greater impact locally and internationally than fighting a phony election with hands and feet bound and mouths taped, and would make it hard for Smith to claim, as he did, 100 percent support of Whites as shown by all fifty seats won.

We agreed to meet early in the morning at the RBC building in Salisbury's Borrowdale suburb. However, upon assembling, it quickly became apparent that I was the only man in uniform, though one other fellow had his camouflage cap poking out of his pocket! Disappointed, I simply waved them all away—"Thanks you guys but just go home."

It took more than a decade of civil war for me to learn that men will commit acts of bravery that are in reality cover-up for lack of moral courage. We men desire recognition from our peers, and that recognition all too often translates as medals on our chests. Moral courage—that is, doing the right thing from the start, before the applause floods in—is not for the faint of heart; it requires weathering condemnation, abuse, ridicule, and rejection.

One evening Evan Campbell (now our former High Commissioner to London) invited me to speak at a dinner for a group of businessmen who were keeping Rhodesia going despite the sanctions. We met in a Salisbury cellar and had an excellent dinner, after which Campbell turned the podium over to me.

I outlined Rhodesia's position militarily, politically and economically, then opened the floor for discussion. Though several had questions, no one disagreed with my assessment that we were losing an unwinnable war and fast running out of time to negotiate an end to it. When there were no more questions, I stated it was my turn to ask questions. I emphasized that never in our short history had we more desperately needed experienced political opposition, leadership, and funding. I pointed out the impressive experience amongst the group: there were former MPs and ministers in the audience who had served in the cabinets of Prime Ministers Huggins, Todd, Whitehead,[1] and Welensky. Why were they being silent when their country needed them more than ever? Their silence made

[1] Edgar Whitehead is the only political figure on this list who I have not yet mentioned. He served as Southern Rhodesia's prime minister from 1958–1962, after Garfield Todd. He was crucial in the negotiation of the 1961 Constitution.

them conspicuous. A couple of men mumbled reasons, and finally Campbell stood up and admitted, "Allan, there is only one honest answer to your question, and that is that we're all shit scared. Smith controls all our businesses through our foreign currency allocations." A ripple of agreement erupted in the room, everyone exclaiming "Hear, hear" in support of Campbell's transparency.

Yet harmful enemies kept entering the scene. One morning a member of our staff drew my attention to an article written about me in the South African weekly, *The Citizen*. It was the most vitriolic condemnation to date. My first comment was that only a mouthpiece for the apartheid-enforcing South African National "Nat" Party could be so extreme and vicious (the main propaganda advisor of the RF, as I knew, was a South African, Ivor Benson). Sure enough, not many months later it came to light that *The Citizen* was a secret channel of South Africa's government. This confirmed what I had feared: the South African government would not only try to close down my ecological consulting work in their country (I was already banned from setting foot on any university campus there and from advising the government of the Bantustan of Lebowa), but would now do all they could to undermine me politically, knowing that I was against Prime Minister Smith.

On June 5, 1973, a delegation from my own RP visited me in Parliament. The group was led by Strath Brown, one of my most trusted shadow cabinet[2] members. Regrettably, they informed me the party wanted me to step down as president. Their reasoning, of course, was that I was so hated and unpopular across the country; the party could never win a seat with me at the helm.

I had no objection to relinquishing leadership, but I wanted to make sure the RP truly understood the consequences. Smith and the RF party almost never attacked our party; they deliberately focused on attacking and smearing me personally. I warned that the moment the RP replaced me, the focus would shift, and Smith and his party would likely begin attacking the RP more so than its leader.

Brown nodded, validating my concern. He even went so far as to draw parallels to their practices as farmers: to eradicate a troop of raiding baboons, they would first shoot the alpha male, which then made the troop easier to destroy.

Despite Brown's analogy and the group's agreement, they said there were just too many in the RP who insisted I go. I then suggested—to avoid fracturing the party—that I relinquish my leadership, saying I was doing so due to all my commitments, including to my international clients. I would remain the RP's sole MP and leave it up to them to select my replacement as leader.

---

[2] A shadow cabinet is a senior group of opposition spokespeople who, under the leadership of the Leader of the Opposition, form an alternative cabinet to that of the government, and whose members shadow or mirror the positions of each individual member of the Cabinet. I did not have a full shadow cabinet but did have a group of my most trusted people for the main portfolios.

A few days later I was informed that the Rhodesia Party had selected their man: Tim Gibbs, the son of former British-appointed Governor Sir Humphrey Gibbs. Tim and his wife Sue were friends of mine, and I knew him to be a pleasant and popular man. Tim's father Humphrey had officially become Rhodesian many years before, when he moved to Rhodesia from Britain. Humphrey took up farming at Nyamandlovu, near Bulawayo, and had a kinship with my father. I immediately endorsed Tim and stepped down.

All went well for six days. Then on June 11, 1973, I encountered another hurdle. I was giving a political speech at Hartley (now Chegutu), south of Salisbury. My talk included my standard refrain: there would never be a military victory, only a political compromise based on seeing the other person's point of view. In an effort to get the audience to empathize with another perspective, I asked, "What if you had been born a Black Rhodesian, denied the opportunity to vote? Had it been me I would have been a guerrilla fighter, and I need you to understand why I would say that." I had hoped that by posing this scenario, I would create an "aha" moment in their minds.

Alarmingly, the next day, the radio, television, and newspaper outlets screamed a stunning takeaway: "Savory says he would be a terrorist!" Tim Gibbs—knowing this was simply not true but anticipating the press would question him—reached out to me for advice. I told him, "Whatever you do, Tim—stand firm. You know I would never refer to 'terrorists' and you can check with anyone at the meeting what I actually said. You also know the RF-controlled media distort anything I say."

Sadly, Tim's version of "standing firm" was not mine. When the press approached him, he validated the terrorist implications by saying he did not support "what Savory said!" Almost immediately the RP began to split. Whereas some members felt Tim was weak and needed to stand by me, others strongly advised that Tim distance himself. To avoid further splitting the party, I resigned from it and became a parliamentary opposition of one independent MP. The RP now had no seat in Parliament and never would again.

As expected, my resignation did not stop the RF's major smear campaign against me, and they barely recognized the party. Knowing a general election was fast approaching, I went along with everyone's belief that I would stand as an incumbent MP in my Matobo constituency. I had several informants within the RF, including a member of the Matabeleland RF committee who discreetly let me know that RF headquarters was circulating stories about my supposedly "scandalous" private life, with instructions to spread the "news."

But I had other plans. When Smith announced the general election in 1974, I quickly switched to Highlands North, a suburb of Salisbury at the opposite end of the country. I was now living there and stood as an independent. Although I felt the election a farce, I knew a one-man boycott would be useless and if any independent had a chance, I was confident it was me, unincumbered by a party.

To my dismay, however, others surfaced to run in the same constituency. Among them were Diana Mitchell (a Centre Party member), my former mentor Dr. Ahrn Palley, and Pip Grinham (who the RP itself put up against me). Immediately, I asked everyone what the hell we were doing; the odds were already stacked against winning, however if four people were running for the same seat against one RF member, it absolutely guaranteed an RF victory!

Diana proposed polling the constituency to see which of us—Ahrn, Pip, she, or me—had the best chance of securing the seat but then Pip and she agreed to step out of the race (both feeling that I for certain had the best chance as an incumbent). However, Pip's agreement was subject to approval by Tim as RP leader. Tim, however, believed the RP had far greater support without me—and would win more votes countrywide if seen to oppose me—so he vehemently disallowed Pip to withdraw. As a result, Diana decided not to withdraw as well. In the end, only Ahrn withdrew. The outcome of the election was as I predicted: all fifty seats went to the RF. Once more, the RF had total control of Parliament. The Rhodesia Party was finished, as was my time in Parliament.

PART X: THE OPPOSITION

CHAPTER 42

THE NATIONAL UNIFYING FORCE (NUF)

My Second Attempt at Leading the Opposition

1975–1977

My belief is that we must genuinely see the other man's point of view.

I was viewed by some as a leader who was forever "tilting at windmills" (pursuing a futile goal), and perhaps not the sort to lead a political party.[1]

By May of 1976 Rhodesia was deteriorating on all fronts. Both the Rhodesia Party (RP) and the Centre Party were deemed powerless. Nevertheless, diehards from the RP still believed that if I returned, they could bring the party back to life, so much so that they boldly approached me to rejoin. After several discussions with the party hierarchy, including Tim Gibbs, I agreed to return, but would not consider leading them. Instead, I agreed to serve under Gibbs.

No sooner had I returned, however, than talks began about the possibility of merging all the non-RF parties. Might we have a better chance against Smith's Rhodesian Front (RF) if unified? Everyone agreed and so the National Unifying Force (NUF) was born—an amalgamation of mainly the Rhodesia Party and Centre Party, but also inclusive

[1] This cartoon by Vic Mackenzie was published in *The Rhodesia Herald* (December 10, 1976, p. 17).

of some of the right-wing elements from past parties. Our aim—much like that of the Rhodesia Party—was to negotiate a constitution for all Rhodesians, regardless of race or tribe, and bring an end to the war. This agreement by all parties to combine was under one stipulation: that I did not lead the new opposition.

Again, I was completely happy with this and emphasized that I did not want to lead the NUF. Though I was in political limbo—having lost my seat in Parliament—I still had army commitments, and my consulting business was continuing to gain momentum. I now had clients in more than five countries in southern Africa. I was dependent on a forged South African passport to visit them all and couldn't afford the extra scrutiny certain to occur if I led the opposition. My work as an independent scientist remained a priority (and was, in the long term, more important than anything politically). I was now also the father of my third daughter, Sarah, who Janet and I had welcomed into the world a few months earlier. I didn't wish to take on any more responsibilities.

When the time came for the NUF to elect officers, I attended the large gathering solely as an observer. The now-unified parties nominated members for various leadership positions, and as agreed, did not include my name. Inevitably, someone from the floor asked, "Why is Allan Savory not mentioned?" The chairman explained as tactfully as he could that my not leading the combined parties was a condition of some for joining. This only served to raise more voices demanding that the members be able to vote for whomever they wanted, ultimately forcing me to stand at the podium and recite my heavy commitments. This only led to more people insisting my name at least be on the list to add to the options they could choose from. I reluctantly returned to the podium. Having foreseen I might need to say more beyond my existing commitments (as I still had a few loyal followers), I now resorted to notes I had prepared in advance in case such a situation arose. Here is what I said:

> *I appreciate the confidence you show in me. However, I believe that when you do a job it must be well-done, and I would have challenges doing that and must be frank about them.*
>
> *First, I am very busy working internationally as well as here and can't give it my full attention.*
>
> *Second, because I will be away a lot, I could only do the job if we had a close-knit loyal team, working under a different structure. Two vice presidents, as you now have, would not be enough. If I were to lead, I could only do so given a larger team that would allow me to delegate.*
>
> *Third, I have little confidence in you. Would you be prepared to be led from the front? Or would I find myself going over the trench on my own?*

> *Remember, I have consistently said Smith must talk to the leaders of ZAPU and ZANU—and I was brought down by the Rhodesia Party, not the RF, because of that. Remember, I have consistently said we need to see the Black man's point of view. Again, I was brought down by the Rhodesia Party, not the RF, because of that. Remember, in the recent election I lost my seat, again due to the Rhodesia Party, not the RF. What guarantee would I have that you would show me any loyalty, even as I would give you my guarantee of loyalty backed up by my proven record? Please think about this and only consider voting for me if you are genuinely prepared to remain loyal at all times, not just at garden parties.*
>
> *If you voted for me, what would you get? I have always believed that one cannot be all things to all people and lead. You would get Presidential directives from time to time. You would be kept to the aim and not be allowed to vacillate. You would find me totally fair and just. You would find me antagonistic to intrigue, pettiness, and small thinking. You would find me blunt and tactless at times. On certain things you would find me completely intractable. My belief is that we must be ruthlessly practical and realistic. My belief is that we must genuinely see the other man's point of view. My belief is in Western ideals of freedom, private enterprise, and human rights. My belief is that what is morally right is politically right.*
>
> *You have a difficult choice because the others nominated are every bit as able as I am and they are tried and tested. Whomever you elect will have my complete support and loyalty. I would like them on my team and therefore am prepared to serve under them.*

And so, my name went forward. At least everyone could decide for themselves, I sighed.

Be it a blessing or a curse, I was overwhelmingly elected leader (president) of the NUF. And with this came five vice president positions! The five included Nick McNally, a respected sharp-minded lawyer (and later a high court judge); Lance Reynolds, a local businessman; and my old friend Pat Bashford. In June 1977, we officially assumed our roles and kicked into action on all fronts, letting the public and the media know that a new and meaningful opposition had formed to challenge the RF.

Sadly, Rhodesia's youth, both Black and White, had spent the bulk of their lives under RF propaganda, racism, and hopelessness. To top it off, due to sanctions, few had travelled overseas to get to know a different world. I was often asked, when interviewed by foreign journalists, "Allan, why are you even trying under these circumstances?" My reply was always along the lines that, while both our present situation and what I fought for were hopeless, "I would keep a flag of sanity flying until the bitter end, for not all Rhodesians are racist nor scared stupid." And now my leadership of the NUF was that very flag of sanity—reinforcing the belief that it simply was not possible to ensure good governance if not governed fairly.

As I saw it, there were now three major political parties vying for power. There were the two African nationalist parties ZAPU (the primarily Ndebele party lead by Nkomo) and ZANU (the primarily Shona party led by Mugabe). Both were banned in Rhodesia but externally supported by the frontline African States (Botswana, Zambia, and Tanzania), Russia, China, and now even Cuba, which made them even more easily branded as "communist" by the RF. The third party was the RF—the Rhodesian Front party—that had essentially become a personality cult led by Prime Minister Smith claiming to defend the "last bastion of democracy in Africa." (This was code for White supremacy; their accusations of "communist" a reflection of their fear of immediate Black power.) And then, barking at their heels—or perhaps mewing like a kitten—was our newly formed NUF party—a White party representing approximately one-third of the White population, not seeking power but with the mission of ending the war by negotiation with a focus on the mediation of good governance for Rhodesians of every color and tribe, based on no "isms" (racism, communism, capitalism, tribalism, or cultism).

Time was increasingly not on our side; Rhodesia's armed forces were running out of manpower. Smith made puerile claims that his government was winning based on the number of "terrorists" killed by security forces each year; I pointed out in public that those so-claimed "terrorists"—killed over the course of two years—had been replaced in about six hours if we considered the Black Rhodesian birth rate.

A new figure had emerged amidst the turmoil, Bishop Muzorewa[2] of the United Methodist Church in Rhodesia. Born near Umtali in eastern Rhodesia, he had attended a mission school there and then went on to Central College (now Central Methodist University) in Fayette, Missouri in the United States before launching the African National Council (ANC) in 1971. There was much talk of Prime Minister Smith favoring an internal settlement with Bishop Muzorewa, who claimed to be "widely supported" by Blacks of all tribes.

Prime Minister Smith used the RF propaganda team to build the illusion of support for Bishop Muzorewa. This support was backed by Zimbabwe United People's Organization (ZUPO), a party largely comprised of chiefs. Though led by Chief Jeremiah Chirau,[3] ZUPO was ultimately controlled by Smith. The RF, through the

---

[2] Abel Tendekayi Muzorewa, commonly known as Bishop Muzorewa, was consecrated as the United Methodist Church's Bishop of Rhodesia in 1968.

[3] Chief Jeremiah Sikireta Chirau, in 1976, formed the Zimbabwe United People's Organization (ZUPO). He was the one and only leader of the party, which disbanded before the 1980 election.

biased media, publicized a statement by Chirau that the chiefs were the true leaders of the people, quoting him as saying to their people, "We are the Chiefs, and we give you orders." Typical of Smith's tactics, this was only a part of Chirau's speech. I countered this publicly by providing Chirau's full speech, which concluded, "We get our orders from government, and we give them to you."

I knew Bishop Muzorewa from a meeting I had had with him, accompanied by his lawyer Edson Sithole,[4] in 1975. We had attempted talks on a prospective future non-racial, nontribal constitution, but I found Bishop Muzorewa clueless and felt he had little chance of gaining meaningful African support. In fact, after meeting with him, I discouraged our party team from any future meetings with Muzorewa. At the very least, if there were meetings, I stipulated that we must avoid discussing anything major unless Edson Sithole was present. Don't get me wrong—I strongly felt that the sooner Smith negotiated, the better—but Smith needed to negotiate with African Nationalist party leaders Nkomo and Mugabe, who had far more Black support (something I knew from talking with the workers on the farms and ranches I visited as well as in their communities and townships).

By this time, I was certain Prime Minister Smith had been having my phone tapped and my mail opened—likely starting as far back as 1972 when I'd crossed the floor in Parliament. Now I wanted to confirm it. Instead of opening a few envelopes the usual way, I ran a razor along the side of each. I then turned the envelopes over to Roy Welensky—who had been knighted by then—and Sir Roy passed them on to British Intelligence. Back came the confirmation: the envelopes had indeed been opened and resealed with a different glue. Sir Roy also let me know that he was confident Smith had all the required equipment to tap phones as he himself had approved the purchase of the equipment during his time as prime minister of the Federation, more than a decade earlier.

I will never know if it was any of this that led to Edson Sithole's death. In October 1975, I had arranged via phone to meet him privately, away from my home or office. I waited for ages at our agreed-upon meeting place, but Edson never arrived. I eventually gave up and went home. Later I learned that Edson had left his office, had a drink at Meikles Hotel, then departed to meet me. On the way he was abducted by suspected members of the police Special Branch and never heard of again.

---

[4] Edson Furatidzayi Chisingaitwi Sithole was the first Black person in the entire southern and central African region to obtain a Doctor of Laws (LLD) degree from the University of South Africa (UNISA).

CHAPTER 43

"BANKING" ON THE OUTSIDE

Colluding with the CIA, the "Enemy," and Reporters

1976–1978

Most of the American reporters followed the public refrain: there are "good guys" who are anti-communist and White, and "bad guys" who are communist and Black. I was complicating that: a non-communist White leading a White opposition?

A-8 The Washington Star Thursday, October 12, 1978

—Associated Press

Rhodesian Prime Minister Ian Smith (left) listens as Rev. Ndabaningi Sithole, a member of the interim government, responds to reporters' questions at the National Press Club yesterday.

3-Man 'Truth Squad' Attacking Smith's Plan for Rhodesia Rule

By Bernard Gwertzman
Special to The Washington Star

While Prime Minister Ian D. Smith of Rhodesia

television and that the opposition party Savory heads has no way of getting its views across. He also stressed that the constitution being drafted by

An article in one US newspaper[1] carried word of the efforts the National Unifying Force (NUF) made—under my leadership—to put right the stories Prime Minister Smith (*pictured left*) and Ndabaningi Sithole (*pictured right*) were telling Americans about the support their interim government would have among Black Rhodesians.

---

[1] Bernard Gwertzman wrote this article, "The 3-Man 'Truth Squad' Attacking Smith's Plan for Rhodesia Rule" (*The Washington Star*, October 12, 1978). *The Washington Star* was published from 1965 to 1979. *The Washington Post* acquired some of the *Star's* assets, though not its archives.

Violence was increasing at an alarming rate among and inside the African nationalist parties as ZAPU and ZANU vied for influence and support within the Black population. This spurred the United States to get further involved in Rhodesia's politics. Under President Ford, with Secretary of State Henry Kissinger as envoy, the US government pressured South Africa to force Prime Minister Smith to negotiate with African nationalist leaders Nkomo and Mugabe as early as possible (though Smith had sworn he would never enter such a negotiation).

When US elections brought in the Carter administration in 1976, President Carter's Ambassador to the United Nations, Andrew Young, joined by Britain's Secretary of State for Foreign and Commonwealth Affairs, David Owen, visited Rhodesia in hopes of peace talks. On several occasions, I led a delegation of the NUF in meetings with Young and Owen. I liked Young with his straightforward talk and approach, but I did not relate well to Owen. I was sitting next to him during a meeting with our small NUF team one day when he pulled a comb from his pocket, swept it through his long hair, and wondered aloud how history would judge him. Such vanity and concern for his legacy[2] dismayed me at a time when some 30,000 Rhodesians had been killed, and we were soon to see thousands more die. The talks were unsuccessful.

All efforts by the British, Americans, the Rhodesia Party and later the NUF were aimed at bringing Prime Minister Smith to the table to end the war and sanctions as quickly as possible. I felt that the longer Smith held out, the weaker the position would be for all moderate Rhodesians, regardless of race or tribe.

In October 1976, I joined African nationalist leaders in Geneva, Switzerland at the beginning of the abortive Geneva Conference organized by Britain to bring about a series of peace talks. In attendance were, on the ZAPU (primarily Ndebele) side: Joshua Nkomo, Dumiso Dabengwa,[3] and Lookout Masuku;[4] and on the ZANU (primarily Shona) side: Robert Mugabe, Josiah Tongogara,[5] and Ndabaningi Sithole (whom Mugabe had left almost without influence). When I met with just the ZAPU leaders, they spoke Ndebele; when I met with just the ZANU leaders, they spoke Shona. However, when we all met together, everyone generally—and freely—spoke English.

---

[2] Owen's "legacy" did turn out well—as David Anthony Llewellyn Baron Owen, CH, PC, FRCP, he was elevated to the House of Lords, where he basks in glory to this day.

[3] Dumiso Dabengwa served as the head of intelligence for the Zimbabwe People's Revolutionary Army (ZIPRA), the militant wing of the Zimbabwe African People's Union (ZAPU) during the Rhodesian Bush War. The White minority nicknamed him the "Black Russian" because he trained in Moscow, Russia.

[4] Lookout Khalisabantu Vumindaba Masuku commanded the armed forces of ZIPRA.

[5] Josiah Tongogara was a prominent Zimbabwean guerrilla commander during the Rhodesian Bush War, operating from outside of Rhodesia's borders to free the country from White rule. In 1973 he commanded the armed forces of ZANLA (Zimbabwe African National Liberation Army), the militant wing of the Zimbabwe African National Union (ZANU). Tragically he was killed in a suspicious "car accident" on December 26, 1979.

Meeting with Nkomo was easy and relaxed. Nkomo knew of my activities through his second in command, Josiah Chinamano (whom I had met and befriended back in 1974) and from Gilbert Ndlovu (my former manager at Kapombe Farm) who I knew was using part of his wages to support Nkomo. On the other hand, my first meeting with Mugabe and his military commander, Tongogara, was tense. Going up to their room in the elevator I was accompanied by a group of men still dressed in ragged clothes, having just been released from detention by President Kaunda of Zambia to attend the meetings.[6] When I entered their room, neither Mugabe nor Tongogara stood up, nor would they shake hands with me. Instead, they bombarded me with questions. To their benefit, their questions were well-informed, and I answered frankly. They knew a lot about me and held me responsible for, as they said, "80 percent of their casualties." They knew my GATU nickname, Chimurenga, and attributed the formation and training of the SAS, TCU, and Selous Scouts to me. I told them my role was greatly exaggerated but did not deny my participation. They then sought answers, primarily, "Why do you lead a Whites-only party and not a multiracial one?"

I answered succinctly, "Under Smith's constitution, multiracial parties are quickly labeled and dismissed as liberal; to be as effective of an opposition as possible to Smith, it is better to oppose him as a White party." To Mugabe, despite our differences, I also pointed out our commonality: we were perhaps the two men in Rhodesia most hated by Prime Minister Smith and the Rhodesian Front. Once they were satisfied with my response to their interrogation, I said it was time to stop killing one another—we needed to work together to bring the war to a negotiated, honorable end. We shook hands. The next day, Nkomo and Mugabe attempted to draft a joint press statement; I offered my native grasp of English in writing it. From then on, we agreed to collaborate where we could politically, while accepting that military opposition would continue until Prime Minister Smith could be brought to the negotiating table.

Not only were these meetings in Geneva an attempted step toward ending the war, but for me personally they also emphasized the great bond I felt with my Black Rhodesian countrymen—a bond stronger than that I felt with the Europeans I met in Europe. Rhodesians like me were truly African—I was a White African, not a White European. This was reinforced during a chance meeting in Geneva. I was walking at night near my hotel, which was some distance from the conference. It was bitter cold, and I came upon a Black man hunched in his overcoat. As I passed him, I chanced that he was Rhodesian and said, "Good evening, it is very cold here!" in Shona. He was overwhelmed with joy and followed me to my hotel to talk. He was now living in Switzerland and—like me—was homesick. We agreed we had more in common as Rhodesians than with any other people.

[6] President Kaunda, who favored Nkomo, had detained many of Mugabe's forces to stop the violence occurring between their two factions within Zambia, which was damaging Zambian citizens and property.

One evening, shortly after my return from Geneva, I had an unusual visit from an American businessman and his wife. We sat in my study and chatted about his desire to start a business in Rhodesia. All of this appeared strange at the time, given that the world—including the United States—was sanctioning Rhodesia. I of course was dissuasive, pointing out all the obvious obstacles for an American with business pursuits in Africa, but the man still insisted on leaving me his "business card"—a nondescript credential listing a Texas address and phone number. When they departed, Janet, puzzled, asked, "Who are they—and what are they wanting?" I replied that I had no idea, but he was either "an idiot or CIA," and filed away the man's card.

A few months later, in early 1978 during one of my trips to the United States to present on my ecological work at the International Stockman's School in Arizona, I took the opportunity to reconnect with the potentially bogus Texas businessman who had left me his card. He wined and dined me and quickly unveiled his real identity, whereupon he arranged to fly me to CIA headquarters at Langley, Virginia, outside of Washington, D.C. After the obligatory lie detector test (the first time I had ever encountered lie detecting machines) to ensure I was not and had never been a communist, we launched into serious discussion. The CIA clearly wanted military and political information on what was unfolding in our civil war. As I was still an officer in the army, and with nearly all my closest friends in the army, I would not provide any military information. I did, however, agree to share my political opinion, insights, and assessments, hopefully to help both the United States and Britain lure Prime Minister Smith to the negotiating table. I was aware of the risk of cooperating with the CIA; few White Rhodesians would ever understand my intent or motive, and if caught, the death penalty was certain.

The CIA, knowing I would divulge no military intelligence, seemed mainly concerned with gathering information on the magnitude of Black support for African nationalist leaders Nkomo and Mugabe. When the RF would announce, for example, that Rhodesia faced 7,000 "terrorists," I would point out that this was pure propaganda—the number of Blacks against the RF realistically numbered in the millions. Even villagers who channeled information to security forces—and faced a brutal death if caught as suspected "vatengesi" (sellouts)—supported the guerrillas by extending safe harbor to them within their communities. On top of this, the inhumane tactics of the Rhodesian Army—from creating internment camps and planting mine fields to burning villages—turned even passive people against the government.

To remain reasonably fluent in Shona, I had befriended a young schoolteacher with whom I offered housing in exchange for conversing one to two hours a day. He chose the topics we discussed, which in the beginning were general but became increasingly political. From him I gained even more insight into what was going on in the townships and what was being said on the street, and passed the appropriate information on to the CIA.

I had my own motive for working with the CIA, however. I was handling a confidential NUF party fund. Not only was the fund secret, but the source of that money was known only to me (in reality there was no fund; the money came directly from my personal earnings in my consulting engagements outside of Rhodesia). As a Rhodesian under UN sanctions, I had no way of banking outside the country, not to mention the Rhodesian dollar was worthless beyond our borders. I had to trust clients and friends to hold money for me in South Africa, England, and the United States and make it available when needed. I was desperate for this to change—I needed a bank account outside Rhodesia, and then a secure means of transferring US dollars and documents into and out of Rhodesia without tipping off Rhodesian intelligence. The CIA did just that; they opened an account in a Swiss bank with me as the sole signatory and gave me an everyday briefcase that contained two false sides, undetectable unless ripped apart. They also provided me with two safe means of communicating with them that Rhodesian intelligence agents would never detect.

In June of 1978 the Progressive Federal Party (PFP) in South Africa, which was the main parliamentary opposition to apartheid, invited me to speak at a conference in Durban. I agreed to do so but did not wish to enter South African politics in any way or tell them what to do in their extremely volatile nation. I decided to just talk about the mistakes we had made in Rhodesia, both Ian Smith and me, in the hope that South Africa would not repeat them. I started by explaining that following the dissolution of the Federation—when Rhodesia reverted once more to a self-governing colony with Britain unable to interfere in our affairs—we should have had immediate discussions on how to accelerate the advancement of all Rhodesians. If we had done this, we would have been unencumbered by UN sanctions and able to negotiate a constitution for all from a position of great strength. But instead, Ian Smith—feeling slighted that Northern Rhodesia and Nyasaland had been granted full independence—declared Rhodesia's immediate Unilateral Declaration of Independence (UDI), thus prompting Britain to interfere and lead global sanctions to bring us to heel. I talked of my mistakes as leader of the opposition, how I should have fought harder in my own party to prevent us from engaging in rigged undemocratic elections or by-elections, and we should have done this by openly boycotting such elections. Actions like this would have made it more difficult for Smith and his RF party to claim he was defending the "last bastion of democracy in Africa." My talk went well, followed by many questions, as well as a few requests (then and later) for me to move to South Africa and lead their opposition to the National Party.

Back in Rhodesia we were now experiencing escalated incidents of ZAPU's guerrilla army, ZIPRA (penetrating from across the upper Zambezi River and Botswana border on our northwest and western boundaries), and of ZANU's guerilla army, ZANLA (penetrating from across the lower Zambezi and down the eastern boundary with

Mozambique). The violence included ambushes, road mining, and attacks on soft targets like mission stations and isolated farms. Most farm and ranch homesteads were now enclosed in fenced, defendable redoubts with radio connection to neighbors and police. My ranch, Gumela, was becoming impossible to defend and I had to move Tom Orford (who was managing it) and his family into town. Visiting my clients was increasingly difficult, especially when attending to projects on Liebig's Ranch. When I visited our most advanced research site, I could only do so in a mine-proofed, well-armed vehicle.

Garth Owen-Smith[7] (who went on to become an award-winning Namibian conservationist), had moved to Rhodesia and taken the position of Section Manager on Liebig's Ranch (years later, he told me he had moved to Rhodesia specifically to find work under me). I often stayed with Garth, as I did with other section managers, each of whom was responsible for about 250,000 acres. During one of those visits, Garth confided that after his senior herder asked him to visit one of the herds, he was met by an ambush party of well-armed guerrillas. The herder, knowing this was coming, quickly cautioned Garth not to do anything impulsive, putting his hand on Garth's weapon to restrain him. The guerrillas forced them out of the Land Rover at gunpoint, then surprisingly just sat with them in the bush, discussing a few things, including management by the staff. They then let Garth and the herder go on their way.

Almost every African—including Garth's staff—was on the side of the guerrillas, and now at least Garth knew that. I advised Garth to report only to Mike Gawler (Liebig's Ranch Manager)—who I would also talk to—and not to report the incident to the police, as they would not believe that the guerrillas only wanted to talk and had then let him go. The police would instead flag Garth as a conspirator and make his life hell, spending more effort investigating him than looking for the guerrillas. Thus, the matter remained between Garth, Mike, the herder, and me.

Liebig's Ranch eventually had bush airstrips built for me on every section, but my plane could not be brought into the defendable security areas. So, during visits, I would have to guard my plane every night, sleeping poorly underneath it. Having had enough of sleepless nights, I finally decided to simply have the staff—obviously on the side of the guerrillas—guard my plane for me. I then slept, worry-free, in a proper bed. The next morning, I found all well with my plane, and I generously tipped the "guards," saying the money was a "contribution to their party's funds." We knowingly nodded at one another. Such was the nonsensical environment of our unnecessary war; to get a good night's sleep, I would simply engage the "enemy" to guard my plane.

While operating across Rhodesia's border into Mozambique, I mentioned earlier that I had tied a prisoner to me with a length of parachute cord to prevent his escape.

---

[7] Garth Owen-Smith was an award-winning South African-Namibian conservationist. He writes about his work in *An Arid Eden: A Personal Account of Conservation in the Kaokoveld* (South Africa: Jonathan Ball, 2010).

In the moments we were sitting on the ground sharing our meals, we had many a quiet, lengthy discussion during which he gave me "enemy" information—and frequently appealed to come work for me after the war. When I handed him over to a Rhodesian African Rifles battalion, he pleaded to remain with me. I felt terrible as he was led away to an unknown fate. Our civil war was indeed a ridiculous situation.

Prime Minister Smith's road was fast narrowing with US pressure on South Africa to effectively cut Rhodesia's supply lines for fuel, ammunition, and almost all imports. Our needs were still great despite amazing efforts to manufacture our own parts and substitutes, including wine and whisky. A visiting British comedian once entertained us by saying he had just had a lovely dinner at the "Tin of Tomato Hotel" (Monomotapa Hotel) and recommended the wine—but couldn't remember if the best vintage was on Tuesdays or Thursdays.

With their options disappearing, Prime Minister Smith and the RF at last conceded to majority rule—but not with Nkomo and Mugabe at the table. On March 3, 1978, they signed the Internal Settlement—an agreement with Bishop Muzorewa, Ndabaningi Sithole, and Chief Chirau. The agreement promised elections in 1979 for a transitional government that would provide for both enfranchisement of Blacks and protection of White political and economic interests. I knew this was only a ruse for bringing in a puppet government led by Bishop Muzorewa[8] (Sithole had been thrown out of ZANU by Mugabe and was now a political nonentity and Chief Chirau was in the government's pay). Smith would remain in government as Muzorewa's close advisor, pulling the puppet strings. A settlement with these three men was meaningless. Had Muzorewa and the others had the support of most of our population, our civil war would have ended with the settlement. Instead, it only got worse.

Smith hoped the Internal Settlement would secure Britain's and the United States' support in dropping the worldwide sanctions against us so that we could resupply and carry on with the war. Now Smith had to "sell" the settlement to Britain and the United States. In October 1978 he leapt at the chance to do so when—over State Department objections—twenty-seven US Senators invited Smith to Washington, D.C. to discuss the settlement. Smith and Ndabaningi Sithole, who Smith asked to accompany him, were issued the required visas.

Realizing that the settlement was only going to escalate the war—even if supported by the United States—I knew I had to get NUF representation (Nick McNally, Lance Reynolds, and me) to the United States immediately. I flew to Johannesburg, South Africa

[8] Only years later was I to learn of the considerable efforts made by some senior civil servants to warn against this puppet government as a solution. For more detail see the document, *H.R.G. Howman on Provincialisation in Rhodesia 1968–1969 and Rational and Irrational Elements* edited by G.C. Passmore (African Studies Centre, Cambridge Occasional Papers, accessed online).

and went straight to the US Embassy to acquire emergency visas. A man by the name of Bill Tilney issued my visa, indicating he was a collector of flags and badly wanted a Rhodesian flag. I promised to send him one if he could get visas for Nick and Lance, too. On that we shook hands. I managed to get on the same flight to the United States as Smith and Sithole with Nick and Lance arriving shortly after. I paid for our airfares and Washington hotel bills from my "secret fund," but money was tight. I got us a room in a cheap Dupont Circle hotel, and we fed ourselves via the least expensive eats possible—prepackaged items from convenience stores.

Once in Washington, I tried to set up media interviews, however every phone contact with a US reporter led to the same polite brush off—"We'll call you back later." As I was flipping through the hotel phone book, my eye caught the number for the BBC's Washington bureau. The reporter who answered immediately knew who I was and asked, "Where are you?" When I responded that I was in Washington, he exclaimed, "Good God—why didn't I know that?" Within half an hour he was in my hotel room interviewing me. After our interview, I asked him about the lack of interest from the American media. He said most of them followed the public refrain: *there are "good guys" who are anti-communist and White, and "bad guys" who are communist and Black. I was complicating that: a non-communist White leading a White opposition? But Smith represented all White Rhodesians!* To clarify my position would take up half of any television or radio interview, hence American newscasters were likely choosing to ignore me. I thanked the BBC reporter for his help and ceased trying to contact the American media, though the Washington newspapers did eventually carry stories on our visit.

With help from Republican friends, I also set up meetings with US senators and a meeting with the Heritage Foundation, the conservative think tank that had sponsored Smith's visit. That meeting was chaired by a rather welcoming former US Army general. At the first opportunity, I asked the general, "Frankly, at what point will the US army become fully involved in Rhodesia as it did in Vietnam?" The general responded that that would never happen, at any point. "Good," I said, "Then can we talk realities."

I pointed out that if the United States supported Smith's plan for bringing about Black majority rule, it would precipitate a massive escalation of the war because the bulk of our population—those following Nkomo and Mugabe—would not support it. The Rhodesian Army was already running out of manpower from a rapidly dwindling White population (emigration exceeded immigration in our small White community), while the government's opponents had a pool of millions of potential recruits and control of Rhodesia's border with Botswana, Zambia, and Mozambique. Rhodesia now only controlled 20 percent of our total border—the portion we shared with South Africa. Given these realities, US support for what I (and most of the Black population) knew would be a puppet government would, in my mind, come with an obligation to provide troops to help deal with the consequences.

That evening, I left Lance in our hotel room and made an excuse to ask Nick to take a walk with me because I needed to know his deeper views should I be eliminated. As we strolled the streets of Washington, I said that I was now truly worried that Smith might try to eliminate me "extra-judiciously," as he was clearly desperate for US support and saw me as a serious obstacle to his plan to bring in a government led by Bishop Muzorewa. If something were to happen to me, Nick would take over leading the NUF, and I needed to know his stance on Smith's plan. Nick responded honestly enough. He said that although he supported me, he felt I was too hard on Smith. If he were to assume party leadership, he would compromise more with Smith and Muzorewa. I thanked him for his response and left it at that, as I would not try to rule from the grave.

Smith and his delegation returned to Rhodesia from the United States without the American support he had so desperately sought, and Nick and Lance returned to give the NUF party a full report. Though we believed Smith would not have gained US support even if NUF representation had been absent from Washington, we felt we had made the right move in being there. People needed to know that Smith did not represent the only White Rhodesian view, nor Muzorewa the majority Black view—and that supporting Smith and Muzorewa would only escalate the violence.

I remained in Washington for a bit longer as there was one more must-have meeting on my agenda: I needed to make sure the State Department knew my fears and predictions, including my concern that a tribal government in place of Smith's would only continue the theme of ruthless domination. I told department officials that Smith's government was quickly becoming a defeated shell, trying to save face and gain at least some cosmetic White representation in any future Parliament. With Smith on the ropes—and any government led by Bishop Muzorewa unlikely to win support—African nationalist leaders Nkomo and Mugabe would be vying for power, each counting on winning the first and all-defining election. By numbers, if it were truly a one-man-one-vote election, Rhodesia would then be under Shona (Mugabe) rule, which wasn't necessarily a good thing. The only cement that was binding the Shona and Ndebele people was their common opposition to White racist rule. As soon as that cement were to dissolve, tribal conflict and violence would follow. Rhodesia's government would continue to be built on the dominance of one tribe and race—it would indeed be one man, one vote, *one time*. If that happened, any talk of a constitution or democratic elections that might enable moderate people to gain seats would be off the table, and inevitably harsh retaliation against Ndebele and White Rhodesians would follow.

Not only did I share my concerns with officials in the State Department, but I stopped in at the Foreign Office in London, enroute back to Rhodesia. In both Washington and London, I stressed the importance of their involvement; I guaranteed further tribal violence that only the US and British governments could prevent. I said there was going to be one chance to "crack the whip," and the US and British governments held the whip handle. I argued vehemently that, no matter how insignificant the NUF might appear, excluding our White nonracial, nontribal point of view would guarantee further killing and severe retribution, likely against the Ndebele minority.

Including the NUF in settlement talks was the obvious place to start. Smith would oppose NUF representation out of pride, if nothing else, having been frequently told by us that he would be forced to negotiate with Nkomo and Mugabe. And both Nkomo and Mugabe would see no need for anyone else's presence (each only concerned with gaining power and believing they would lead the first Black government). Without the NUF, there would be no party at the table that was seeking an end to the war (rather than immediate power) and demanding a constitution representing the security and future of *all* Rhodesian citizens, regardless of race or tribe. Despite my desperate appeals, I could not rally the slightest interest in preventing what I was predicting. I returned home to Rhodesia, the horror of what was to come weighing heavily on my mind.

I followed up with my promise to send Bill Tilney (the issuer of our visas) a rare Rhodesian flag. Years later, after Bill had retired, I met up with him. He confided that he had kept a copy of my visa applications and other papers concerning me because he felt they would one day be valuable—a value I question, as my failure to persuade the US or British governments to address my concerns indeed resulted in bloodshed.

CHAPTER 44

TRY SAVORY FOR TREASON

Escape

1978–1979

Smith's government would soon act to silence me as they had Edson Sithole and others.

Saying goodbye to my children in Durban on my escape into exile. I'm holding Claire; Rodger stands in front of Megan. Behind us is my plane, "Strela-painted" grey to foil heat-seeking missiles (1979).

In October 1978, on my return to Salisbury from London, Janet met me at the airport. Driving home, we passed large posters lining the road with the message, "Try Savory for Treason." In classic Smith style, the prime minister was seeking retribution for his failure in Washington. In Smith's opinion, my suggestion that he should talk to "terrorists and communists" was an irresponsibility that went beyond reckoning. "There is no end to Savory's irresponsibility!" Smith declared. "As a man of principle, upholding the last bastion of democracy in Africa, I would never do what Savory did, ever!"[1]

Within a month of being back, Foreign Minister P.K. van der Byl[2] and others in Smith's cabinet were publicly calling me a traitor, as they had many times before. Like

[1] Ironically, Smith ends up doing just this—talking with "terrorists and communists"—details of which I share in the next chapter.

[2] Pieter Kenyon Fleming-Voltelyn van der Byl served as foreign minister of Rhodesia from 1974 to 1979. He was one of the leading agitators for Rhodesia's Unilateral Declaration of Independence (UDI) in 1965 and was responsible for introducing press censorship.

the posters that had greeted me on the drive home, they demanded yet again that I be tried for treason for my insistence that they bring all parties to the table for negotiation. (While still leading the Rhodesia Party two years earlier, I had responded to these same charges publicly through press interviews, stating, "I welcome such a trial, as it would give Rhodesians a chance to hear what I have to say without my words first being filtered by the Rhodesian Front." My public statements made it into a few of the local newspapers but of course never made it into the heavily censored radio and television broadcasts.)

In 1979 journalist David Usher conducted an interview with me that was published in a Durban newspaper, the *Sunday Tribune,* on February 25 under the headline, "The Guerrilla War Smithy Just Can't Win." The subheading (which was of course blocked out in Rhodesia) read, "If I was Nkomo, the first thing I would pray for would be that they would be stupid enough to call us terrorists." It was among the hardest-hitting interviews I had given, knowing it would be published in South Africa—the country keeping Smith in power.

All along, South Africa's support had been essential to Rhodesia's survival, and the apartheid-enforcing South African National ("Nat") Party had ensured this support, although South African businessmen were charging us astronomical prices for their wares (because they could). Rhodesia's main RF propaganda advisor, Ivor Benson, was South African; South Africa allowed fuel to flow to Rhodesia through its port in Durban; and South Africa supplied ammunition and aircraft parts, loaned helicopters, shared intelligence and much more. With the Zambian border now closed, evading sanctions became at best extremely difficult. Now everything coming out of Rhodesia was clearly Rhodesian and thus could not leave South African ports without the blessing of their government, and everything coming into Rhodesia was not headed to any other country. Checkmate.

Unknown to me at the time, on the same day the *Sunday Tribune* had published a second article headed, "Search Warrant After Interview with Anti-Smith Man. Police Seize Savory Tapes from *Tribune*." The article reported that the police (South African Special Branch), accompanied by a Detective Inspector of the Rhodesian Criminal Investigation Department (CID), had gone to the newspaper's Durban offices and demanded they turn over the taped record of my interview with Usher.

Back in Rhodesia, however, I didn't encounter hostility to my interview until Sunday, April 15. That evening, at my home in Salisbury, I was walking in my garden when a vehicle drove up with two police officers from Special Branch, obviously on official business. I invited them into my study, where they sat together on a couch opposite my cushioned chair. One of them handed me a copy of the *Sunday Tribune* article in which quotes of mine had been heavily underlined.

The officer who I assumed was senior asked, "Did you make these comments or were you misquoted?" I replied affirmatively—I had said what was underlined. In fact, I emphasized that I had been accurately and clearly quoted—did they feel my statements were untrue? At this point, the other policeman leaned back as far as he could on the couch, so his face could not be seen by his colleague. He silently mouthed a warning that I was in danger. As subtly as I could, I acknowledged I understood, relieved that at least one of the two policemen didn't blindly follow the Rhodesian Front. I then asked, "Who underlined my words in this article?" and learned that it was Rhodesia's Attorney General, Hilary Squires, who I knew. I asked them to give the Attorney General my regards and tell him that it is "sick minds like his who are endangering our country." The two policemen promptly thanked me and left.

Shortly before the previous encounter, I had received a warning from a British source letting me know that a deserter from the Rhodesian Army was in England. The deserter had apparently been interrogated and had passed a lie detector test, and they wanted to verify part of his story. The deserter claimed that he had been trained by me; of most alarm, however, he had seen my name on an assassination list. I didn't know the man and told my source that despite passing the lie detector test, the deserter was lying. I explained that often, men falsely claimed to be trained by me, or to know me. On one occasion even, a man was speaking to *me*, embellishing on how well he knew me, until I eventually introduced myself!

That said, the deserter's claim, combined with the quiet warning from my unknown Special Branch police friend, made me certain Prime Minister Smith's government would soon act to silence me as they had Edson Sithole and others. Consequently, I anticipated how Smith (and South Africa's government, which would certainly act on Smith's behalf) might eliminate me. Those in the army and police that I'd trained or had operated with remained loyal to me and would challenge a government claim that I had simply "disappeared." The most likely move to get rid of me, therefore, would be through detention as a traitor, which the RF propaganda over the years had already assured many people I was. Thus, when the Special Branch policemen had come to my home to question me, I took the opportunity to provoke the Attorney General into acting now, rather than at a time when I might be caught off guard. Once the policemen reported my responses to their questions, it would only be a matter of days before I was detained on the charge that I was publicly undermining Rhodesia in a foreign country (South Africa), clearly a traitorous act. In detention I'd "slip" on soap in the shower and break my neck, or accidentally "fall" out of a window.

But I was prepared. During my visits to Britain and the United States I had scouted possible places I could escape to—should the need arise—and continue my scientific work, supported by my ecological consulting. After visiting both the British Channel Islands off the coast of France and the Cayman Islands in the Caribbean, I decided on the latter. I could work in both North and South America more easily from the Caribbean (and the fishing was better!).

That night I let Janet know that I had no option but to escape at dawn (the earliest I could depart without raising suspicion, allowing me time to set a false trail). I would find my way to the Cayman Islands where she and our daughter Sarah (now nearly three), could join me as soon as I was settled. I assured Janet that I was the one in danger, not her—and that the war couldn't possibly go on much longer.

To quell any suspicion that I might run, I quickly made false plans for the following day. Knowing my phone was tapped, I called both NUF vice presidents Nick McNally and Lance Reynolds and set up meetings with them over the next week.

I needed to get to Lusaka, Zambia or Gaborone, Botswana, but in Rhodesia every civilian flight had to have an approved flight plan before takeoff, and flying directly to either would invite interception, as would deviating off course to either destination once airborne. I would have to flight plan and fly first to the only place unlikely to raise suspicion: South Africa. If the government were suspicious, it would simply have its South African counterparts arrest me, rather than risk shooting me down.

I worked out a low-level evasion strategy, just in case my departure aroused suspicion after I was airborne and on radar, and the Rhodesian authorities decided to intervene rather than rely on the South African Special Branch. Although my plane, unlike all other civilian planes, had Strela paint to foil heat-seeking missiles, that was of little use in evading the Rhodesian Air Force. I would have to fly at tree-top level, something I was abnormally experienced at.

Not knowing when I would be able to return home, I said goodnight and goodbye to my daughter Sarah, packed a few items of sentimental value, tucked important papers and money into the secret compartments of my briefcase, and attempted to get some sleep.

I kept my plane fueled and standing by at Rhodesia United Air Carriers (RUAC) facilities on the civilian side of the main field opposite the air force base. At dawn Janet drove me to the airport, dropping me off by the control tower where I filed a flight plan to Wonderboom, an airfield north of Pretoria. I made a point of chatting with the air traffic controllers who I knew well, casually saying I was going down to South Africa for a meeting and would be back the next day.

With flight plan approved, I departed, climbed to my assigned altitude, and set course. At each reporting point I checked in as required and had an uneventful flight. After landing at Wonderboom and clearing customs and immigration, I phoned Stan Parsons (the agricultural economist who had been part of the Charter Estate Grazing Trial) and stayed with him that night to hand over the management of my South African and Namibian clients. I also called my ex-wife, Shirley, asking her to meet me the next morning with the children at Virginia airfield, a small grass airstrip on the seashore near Durban. On arrival at Virginia, Shirley was there with Megan, Rodger, and Claire (Jim had joined the South African army and was away in Angola).

Upon understanding what was happening, Meg asked, "Dad, did you take the photo albums?" The photo albums were indeed in my possession. The evening before, while wandering around the house trying to think what I might take into exile on a distant island, I realized almost nothing had real value to my family other than our memories. Thankfully, there was just enough room in my luggage for our photo albums, filled with black-and-white pictures of so many happy times. I told the kids that I didn't know when I would see them again, but that I would write from the Cayman Islands and have them visit as soon as I could. And then, with an aching heart, I hugged them all goodbye, handed my 9mm pistol to Shirley (as I would have no further need for it), and quickly left.

To cause confusion in case my movements were being monitored, I then flew to Margate and Pietermaritzburg before returning to Wonderboom. From there, as I was now flying out of the country, I filed a flight plan to Mbabane, the capital of Swaziland (a small country entirely enclosed by South Africa and Mozambique and now called Eswatini) and took off. Once my plane climbed to altitude and was clearly on course for Mbabane, I turned it in a tight circle, descending to treetop level (knowing I would be on South Africa's radar and if observed, would appear to have crashed). I then changed course from northeast to northwest. Maintaining radio silence, I flew on at treetop level, below radar, to Botswana. Once I crossed the border out of South Africa, I climbed to a higher altitude and landed in Gaborone.

I phoned Janet to let her know I was now safe from both the Rhodesian and South African police, then asked her to convey my escape to Nick, Lance, and the rest of the NUF leadership and tell Nick to take over. Janet also arranged for a friend in the air force to collect my plane and store it for me. I then phoned the British embassy in Lusaka, let them know what had happened, and took the next commercial flight to Lusaka.

As our plane approached the landing at Lusaka, nervous Zambian troops opened fire—which I was only aware of because tracer bullets passing by caused our pilot to take evasive measures—alas, a sign of the times. In Lusaka I stayed with my friend Guy Scott.[3] No sooner had Guy and I finished dinner than the phone rang. It was James McMillan, the editor of the *Natal Mercury*, calling from Durban to warn me not to come into South Africa. James had just been visited by the South African Special Branch and asked about my whereabouts. He had immediately phoned Janet to warn me, and Janet had given him Guy's number. I had more than once wondered if I was being overcautious and dramatic in my need to escape but now was grateful that I had done so.[4]

The next day in Lusaka I met privately with ZANU leader Mugabe, who was staying in Zambia as a guest of President Kaunda. We had a friendly talk in his bedroom,

[3] Guy Lindsay Scott ended up serving as vice president of Zambia from 2011 to 2014, then acting president of Zambia from 2014 to 2015.

[4] A senior Rhodesian Special Branch officer later confirmed that I had indeed been on an assassination list.

discussing, amongst other issues, my political future. I informed Mugabe that I would now be out of the picture (it was of most importance to engage full-time in my ecological work). Mugabe also asked my views on various senior army commanders, which was interesting. As we parted, I mused that I expected him to win any future elections in Rhodesia and wished him well with the hope that he could unite our country.

While in Zambia I also met with ZAPU's leader Nkomo, but it took some time to find him. When I arrived at the house that was being used as ZAPU headquarters, I was told to talk first with Nkomo's military commanders, which I did. My meeting with the ZIPRA commanders was a friendly but strange encounter with few questions asked. I did have one objective, and that was to persuade them *not* to involve the Cubans nor their armored vehicles because they didn't need to; the war would end soon if they just kept on as they were. It was also unwise; the Rhodesian Army—whose infantry and special forces were superior in Rhodesia's dense bush—would quickly destroy armored cars and tanks. Not to mention, when the war inevitably ended, a direct Cuban presence would add an unnecessary complexity.

The military commanders had me stay at their headquarters house until someone could take me to Nkomo, who for security reasons stayed elsewhere. During the rather long wait, I chatted with the soldiers guarding the house and—being irreverent—teased them about the sloppiness of their defense. Assessing their positions in relation to the many approaches to the house, I pointed out that if we were to raid them, they would be wiped out with little trouble. They assured me, laughing, that that would not happen—they were the "best."

Midafternoon I was escorted to a house a couple of miles away and had a private meeting with Nkomo. His commander, Dumiso Dabengwa, had already informed him of my views regarding Cuban involvement, so now I focused on my main concern: the tribal conflict I predicted would happen once the Shona presumably took power. Nkomo was confident he would win any elections and would thus personally ensure stability. He assured me that—contrary to my beliefs—their parties and armies were not tribal, and he could foresee no conflict. I could only continue to warn him; he would *not* win because Shonas would vote *en masse* for Mugabe—and ZANU and their ZANLA armed gangs had far wider influence among the population than ZAPU and its ZIPRA combatants. I asked what possibility there might be for ZAPU and ZANU to unite against the Smith-sponsored Muzorewa group but that, too, Nkomo rejected as unnecessary.

In the hope that the NUF would be present at the constitutional conference following the end of the war, I had drafted a plan for a voting system that would ensure extremists of any tribe would be unlikely to govern. The system enforced not only "one citizen, one vote" but made it so candidates—to get elected—had to win a certain percentage of their total votes from races and tribes other than their own. Nkomo agreed that such an unprecedented plan might indeed favor election of moderates, but once more assured me there was no need for this; he would win, and the Ndebeles would not clash with the Shonas under his leadership.

Before departing I urged Nkomo to at least do his best to see that the moderate Whites represented by the NUF were participants in any settlement talks. Nkomo's advocacy was critical as Prime Minister Smith would surely oppose NUF's presence and Mugabe would dismiss the NUF as a nuisance. I also had little faith that the cynical British or naïve US governments would think it important to have the White opposition at any talks—nor to raise the issue of post-settlement strife at all. Nkomo wrapped one of his enormous arms around my shoulder and told me not to worry. He was sure any settlement talks would go well, and he would prevail. Nevertheless, I gave Nkomo my handwritten draft of my voting system plan. I was never to see Nkomo again.

I departed for London that evening. I had phoned the British embassy, explaining my new status as a refugee in limbo trying to get to the Cayman Islands and stressing my need for a more secure passport than my forged South African one. They instructed me to go to the side door of a certain building in London, and all would be fixed.

Upon my arrival in London, I was shocked to learn over the news that during the night, while I was in the air, the Rhodesian SAS had raided Nkomo's headquarters and killed all the guards with whom I had joked. Nkomo's life was spared, only because he was sleeping elsewhere. I nonetheless worried that Nkomo would link the raid to my visit. I called to assure him that I had known nothing of the raid—if I had been involved, the SAS would have known where to find him. Nkomo accepted my logic. Sometime later, Colonel Brian Robinson, my friend and former commander of the SAS, came to stay with me in the Cayman Islands. Brian let me know that the SAS had confiscated notes that the ZIPRA commanders had taken during my meeting at their headquarters. The Muzorewa–Smith government had made great use of the ZIPRA notes, further portraying me as a traitor.

As the British embassy had promised, I was admitted at the side door of the passport building and in no time at all had a British passport but without the right to reside in the United Kingdom. The story of my turbulent military and political life ends at this point when, at the ripe age of forty-three, I landed on Grand Cayman, a Caribbean Island 8,000 miles from the place of my birth.

CHAPTER 45

A SPECTATOR FROM AFAR

How It Ended . . . or Perhaps Started

1979–1980

[Allan Savory] was inclined to be right at the wrong time.

Janet and Sarah near Salisbury (now Harare) shortly after we formed the NUF (1977).

Grand Cayman is as strong a symbol of exile as ever was in our modern world. Once a pirate haven, it is now a tax haven, its shores fished by the descendants of renegades and slaves, its streets home to stateless capital. More than 122,000 companies were registered in the Cayman Islands as of 2024, including more than 100 banks, 700 insurers, nearly 13,000 mutual funds, and a stock exchange. The highest point on Grand Cayman rises only eighty feet above high tide. That is the trash dump.

I fit the profile of a Cayman refugee. I had lost all welcome in any country (though the British had given me a passport, it was without the right to reside in the United Kingdom) and my only tangible resources were the few I had managed to sequester abroad. As a barefoot fisherman on a Caribbean Island, I was powerless to do more for my country. I was now simply a spectator from afar.

Back in Rhodesia, when the NUF announced that I had resigned as president, a leader (editorial) appeared in the *Bulawayo Chronicle* on April 11, 1979, reading:

Right and Wrong

> *The turbulent political career of Allan Savory, now in cold storage because he has resigned as President of the NUF, offers a fascinating study in the ironies of public life.*
>
> *He was inclined to be right at the wrong time. His arch enemy in politics, Ian Smith, held power and prominence by being wrong at the right time.*
>
> *Mr. Smith kept on being wrong so successfully that when finally forced to admit it, he lost scarcely a vote. Mr. Savory has been right so unsuccessfully that, even though Mr. Smith has proved him right, he would win scarcely a vote.*
>
> *Will our future rulers be Smiths or Savorys?*
>
> *Smiths, of course, or they won't rule. A chilling thought.*

And on April 19, NUF's new president, Nick McNally, circulated the following letter to his party members:

> *I would like to pay warm tribute to [Allan Savory's] courageous, albeit controversial political stand over the years. He has made his mark on the history of this country. I believe he will be seen in the years to come as the man who stood out consistently against the popular tide, and who proved right time and time again. Yet no sooner were people grudgingly admitting that perhaps he was right last time, than he was off again uttering fresh heresies and shocking them all over again. My own political approach is a great deal more mundane and a great deal less controversial.*

Nick then laid out the changes to be made under his leadership as president. This involved taking a conciliatory line with Prime Minister Smith, just as he had so honestly told me he would do during our evening walk in Washington, D.C. Still, it came as a shock to me when he announced the NUF's support for Smith's Internal Settlement.

When the elections stipulated in the Internal Settlement were held in April 1979, Nkomo and Mugabe (and their respective parties ZAPU and ZANU) refused to participate, so the only people who voted were those who supported Smith, Muzorewa, Chirau, and Sithole. The United Nations declared the results null and void—the election was not representative of the people.

The following month—May 1979—was an active one. In Britain, the general election on May 3 brought Margaret Thatcher's Conservative Party to power. Then, on May 21, the US Senate overwhelmingly voted on a resolution recommending the

Carter administration support Smith's Internal Settlement. Smith and his RF party, after years of bravado and bringing up a whole generation of youth under racist propaganda, surrendered power to Bishop Muzorewa, a man I had judged too unintelligent to even discuss a future constitution. On May 30, the press reported the swearing in of Muzorewa as prime minister and on the next day Muzorewa appointed Smith to his cabinet as minister without portfolio, which enabled him to devote his time to advising Muzorewa.

When Nick backed Smith and Muzorewa, the NUF was politically neutered. Now no one represented the views of moderate (nonracist, nontribal) Rhodesians, both Black and White. Britain's stand—that there had to be all-party settlement talks leading to peace and lifting of sanctions (which I had consistently favored)—now wobbled. Rhodesia's name was changed to "Zimbabwe-Rhodesia"—a short-lived name under a short-lived puppet government.

All African states—including neighboring states bearing the brunt of Rhodesia's war—refused to accept the legitimacy of the Muzorewa government. This harsh reality, combined with both Britain's and the United States' desire for trade with African nations (in particular, access to Nigeria's huge oil reserves), led them also to refuse to acknowledge the new Muzorewa government. While the British and US governments backed off for the wrong reasons, it was the right action. To have supported Muzorewa and Smith would have prolonged the war, resulting in the loss of many more lives.

Britain and the United States' refusal to accept the new government hammered the final nail into Smith's political coffin. Nine months later, in December 1979, the settlement known as the Lancaster House Agreement[1] was arranged by Britian and signed, forcing Smith to end the war and agree to hold another general election. Tragically, with Nick having thrown NUF support behind the Muzorewa-Smith government, this was not the all-party agreement it should have been. The three parties invited to the talks (Mugabe's ZANU, Nkomo's ZAPU, and Muzorewa's new government) were—as I had feared and predicted—only negotiating a hasty end to war and sanctions, each believing their party would gain power. The White Rhodesians likely to remain in the country—as well as the minority African tribes and the Colored (mixed race) and Asian (Indian and Pakistani) Rhodesians who sought good governance rather than political power—had no voice at all.

Zimbabwe-Rhodesia reverted briefly to British rule and this time the general election—held in February 1980—was not boycotted by Nkomo and Mugabe and their parties. As I had predicted, ZANU, now ZANU-PF (for Patriotic Front) won by a landslide and Mugabe was elected prime minister (this was despite evidence of large-scale voter intimidation by his party, which should have disqualified him). On April 18, 1980,

[1] The Lancaster House Agreement, signed on December 21, 1979, effectively concluded the Rhodesian Bush War and permitted ZANU and ZAPU to stand candidates in the forthcoming elections.

independence was granted to Zimbabwe-Rhodesia and the country became Zimbabwe; we were now all Zimbabweans.

When Smith was forced to end the war, he did what I had hoped he would have done much earlier (and I was accused of being a traitor for suggesting): talk with so-called "terrorists and communists." As Smith notes in his memoir:

> *On our second evening in London we attended a social get-together to enable delegates to meet. A tall fine-looking black gentleman approached me and shook my hand warmly, saying: "For many years one of the things I hoped for was that we could work together in order to build our country, instead of killing one another. This is a happy occasion." I asked his name. "Josiah Tongogara, he replied."*[2]

And after the first announcements of results from the election that followed, Smith wrote:

> *At 7 p.m. I received a phone call to ask if I would visit Mugabe at his house in Mount Pleasant. I was welcomed most courteously, and Mugabe ushered me to a seat in his lounge. He said the results now indicated that his party had won a majority, and I was the first person he had called in for an exchange of views.*[3]

Smith also recounts a civil discussion with Mugabe about the need to keep the country stable and productive for everyone. He then muses:

> *When I got back home, I said to Janet [Smith's wife] that I hoped it was not a hallucination. He [Mugabe] behaved like a balanced, civilized Westerner, the antithesis of the communist gangster I had expected. If this was a true picture, then there could be hope instead of despair. But it would be wise not to jump to conclusions because communists are cunning tacticians, noted for their skills in psychological warfare.*[4]

Tragically, by the time Smith was talking to "terrorists," it was too late to agree on a civilized constitution; White Rhodesians and Ndebeles now found themselves at the mercy of a Shona-ruled government.

Initially, Mugabe attempted to govern well, taking unexpected actions including appointing Lieutenant General Peter Walls to command the new Zimbabwe Army, which included men selected from the ZIPRA, ZANLA, and Rhodesian armies. (This

---

[2] *The Great Betrayal: The Memoirs of Africa's Most Controversial Leader* by Ian Douglas Smith (London: Blake Publishing Ltd., 1997), p. 314.

[3] Ibid., p. 341.

[4] Ibid., p. 342.

explained why Mugabe had questioned me on the various senior officers, including Walls.) Mugabe also appointed Nkomo to his cabinet as well as a few White former ministers, including former Minister of Agriculture David Smith as his minister of finance. Reconciliation was in the air. From my island haven, I wrote a letter to Mugabe wishing him well, and then another to Lieutenant General Walls, apologizing for having had to oppose his actions as strongly as I did, and also wishing him well in his new role.

In June 1979, Janet and Sarah joined me on Grand Cayman. Free at last from the awful stresses of politics and war, I fell into a rhythm of excursions to my clients in North and South America, punctuated by lazy weeks with Janet and Sarah relaxing on the island. Frequently Janet pleaded with me to give her a date when we could return to Zimbabwe. I make no promises I cannot keep and so only committed to returning as soon as I could legally do so. I too wanted to be home, where I still had two of my own ranches (Gumela and Waterford), the bush and the wildlife I loved, and lifelong friends.

A day came when, upon returning from one of my work trips, I found my home empty—with only a single pair of Sarah's tiny shoes to show that my wife and daughter had ever lived there.

Too late, I realized it was unfair of me to have expected Janet to live as we were doing. She had always been surrounded by family and close friends and couldn't bear to live in relative isolation on a small island. Nor could she adjust to life with a "nobody" of a husband when she had been used to one constantly in the limelight. I recognized that Janet, raised in a deeply political family, had married me largely for my prominent stand against Smith. Now she was experiencing a more authentic version of me—one with no desire for recognition, status, celebrity, or power, but rather a driving desire to solve what I increasingly recognized as one of the world's greatest problems: the accelerating destruction of our environment inevitably leading to social breakdown, violence, and war.

Unable to follow her—and after lying four days in a deep depression listening to classical music and barely eating—I knew I had to simply get on with life and bury my sorrows in work, which I did.

I probably wasn't the only beach comber in the Cayman Islands who believed that his insight might change the world, but while that book is still being written, I must close this one with a dash of hindsight on how exile, while unfortunate for me, led to an extraordinary breakthrough. My four-year exile became the most creative and productive period of my life, enabling me to understand that desertification was only the tip of a far greater iceberg—that of climate change, which it was exacerbating in high measure—and what we could do to successfully address it.

EPILOGUE

World leaders continue to believe that fossil fuels and livestock cause desertification and climate change, not our management of them, and their policies reflect it. That leaves future generations facing disasters we cannot begin to imagine, if we don't change this very soon.

(*top*) Jody and me as I'm receiving Australia's International Banksia Award (for the "person doing the most for the environment on a global scale," 2003). (*bottom*) One of the last groups to take the US government training courses I ran under contract. I'm seated in the front in the striped jersey with Bob Vaughan-Evans (to the right of me), who had joined us from Zimbabwe (1997).

Although amnesty was granted to all Zimbabweans in 1980, I found I could not return because the "secret fund" I'd established outside the country to finance the overseas efforts of the NUF was considered an illegal action in a still-closed economy (and was not covered in the amnesty). With the help of Sir Garfield Todd, Prime Minister Mugabe, and my lawyer, Lindsay Cook, a way was eventually found for me to come home and not be automatically detained. Prime Minister Mugabe gave me a pardon and guarantee of safety, and I returned to my country for the first time after four years in exile. Upon arrival, I was nevertheless pulled out of the queue, and security men confiscated my passport. I suggested they talk to the prime minister before any further action and this they did. Mugabe honored his word to me, and they returned my passport that evening.

Unfortunately, under Prime Minister Mugabe's rule, what I had feared and predicted for my country unfolded. The Shona–Ndebele tribal split soon became obvious when retribution against the Ndebeles followed. From 1983 to 1987, an estimated 20,000 unarmed men, women, and children—mostly Ndebele—were slaughtered, their bodies dumped down disused mineshafts.[1] Most of these people were killed by the all-Shona army unit known as the 5th Brigade—a unit trained by North Koreans (at Mugabe's request) to purge the country of "dissidents."

When I had visited the US State Department in Washington and the British Foreign Office in London in 1979, warning of the troubles to come, the State Department officials had at best been grossly naïve. They had assured me that everyone would be sick of the war and unlikely to wage another once Smith's government had stepped down. In contrast, their counterparts in the British Foreign Office—with centuries of experience behind them—simply said, "We believe you, Allan, but it won't be on our plate."

The United States, Britain, and the rest of the world turned a blind eye to the genocide. To this day *Gukurahundi* ("washing away the chaff" as this genocide was called in Shona) remains a festering sore. (Its leading architect, Emmerson Mnangagwa—or The Crocodile, as endorsed by the wearing of Lacoste shirts with the crocodile logo—replaced Mugabe as president of Zimbabwe in a *coup d'etat* in 2017.)

Despite Prime Minister Mugabe extending an olive branch, the White population plummeted from more than 200,000 to about 50,000 as droves, including farmers, left the country. Of the Whites remaining, many of the younger ones had high hopes for the future, a deep love of Africa, and were reluctant to leave. They gladly accepted Mugabe's olive branch but would not consider buying land from the many departing farmers until Mugabe's government issued "Certificates of No Interest"[2] in settling Black Zimbabweans

[1] More of Rhodesia's own people would be killed by the new Zimbabwean Army than foreign enemies killed by the Rhodesian Army through the Boer War, two World Wars, and the Malayan Emergency combined.

[2] A "Certificate of No Interest" was a legal assurance by the government that the land was not required to settle people on and could therefore be purchased.

on that land. Thereafter, young White farmers accounted for 80 percent of the many farms bought after the war.

By 1984, Zimbabwe's former world-class agriculture was once more forging ahead and the country became a food-exporting nation. Unfortunately, Prime Minister Mugabe's guarantee of "no interest" was not to last. The concept of a nonpolitical professional civil service was replaced by one more familiar to African cultures based on tribal affiliation and nepotism. The numbers of senior, unskilled civil servants boomed, and inefficiency increased. Escalating unemployment and mounting criticism of the government led to the rise of political opposition and a new political party in 1999, the Movement for Democratic Change (MDC). Mugabe—whose increasingly incompetent government could not spur economic growth nor curtail unemployment—offered land instead of jobs as a distraction (a common political ploy). The offer of free land combined with political demagoguery rallied masses of people to claim that land had been "stolen" from them by the White farmers, many of whom were MDC supporters. Hordes of people proceeded to invade the more than 4,800 farms owned by Whites, killing some farmers who resisted. This farm takeover also included farms of a few non-ZANU-PF supporting Blacks. Predictably, the economy began to collapse and to address it, the government printed money, resulting in the world's highest-ever inflation rate. I am reminded of this today as I look at the 100-trillion-dollar notes in my possession (three zeros had to be removed because accounting software couldn't handle them all).

To add to the tragedies, the "one man, one vote, one time" prediction became a reality. The ZANU-PF party has held power for more than forty years. No election that would meet the requirements of a *democratic* election has been held since 1962 (when Winston Field was elected prime minister). Ian Smith's internal coup upset even that one, and Rhodesia-now-Zimbabwe has been governed for more than half a century by people (first White and then Black) hanging onto power by suppressing dissent—often violently—and manipulating elections. Through it all, army and police officers have enabled the oppression of their own people—including their own families—by violating their oath to support the constitution in favor of loyalty to party, as in the days of Ian Smith.

As concerned as I continued to be for my native country, but knowing I could do no more politically, I now turned my focus back to the greater problem of desertification and how we could resolve it. I had shown in the Charter Estate Grazing Trial that we could reverse the degradation using livestock in a way that earned rather than cost money. And I also had the data from the two Advanced Projects to back me up. Arizona rancher Ted Spurlock and a few Texas ranchers helped secure a US visa for me so I could now take them on as clients. Liebig's Extract of Meat Company (which owned Liebig's Ranch) was still engaging me on their properties in Paraguay and thus I had clients in both North and South America. Within a year I managed to buy another plane, this time in the United States, to help me travel from client to client within North America.

In January 1980, while passing through the United States, I was interviewed by a freelance journalist in New Mexico named Jody Butterfield. She was unusual in that she had done a great amount of homework, visited some of my ranching clients, and read numerous books on desertification and its ruinous effects on past civilizations. We exchanged addresses and corresponded for months. I had by then moved from Grand Cayman to Cayman Brac, where I'd bought a modest house. Frequently I pedaled my bike the several miles to the island's post office, eager to see if there was a letter from Jody. Entirely through correspondence, we acknowledged a mutual kindling of affection. Six months later we decided to meet again. Two years later, the local preacher on Cayman Brac married us. For more than forty years, Jody and I have been devoted partners in our quest to turn the discoveries from the bushveld of Southern Africa into a hopeful revelation for our planet. For this schoolboy-who-failed-English, Jody's journalistic skills have helped immensely in refining my ideas to the public.

For all its success, the significance of my work also garnered attention in unfortunate ways, among them plagiarism. At one point I was handed promotional pamphlets, published in 1978, that described short-duration grazing (as I was calling my work in Rhodesia by then), claiming it had been developed in Texas! The scholars plagiarizing my work had not fully understood it and promoted it as a preformulated, short duration grazing system. They had entirely glossed over the essential planning process. Rather than fight plagiarism by a powerful university, I changed the name from "short duration grazing" to "Savory Grazing Method," addressing the stark difference between a grazing system that could only succeed if everything was entirely predictable and a planning process that could only fail if not used.

When working in the United States, I lived inexpensively on the "wrong side of the tracks" in San Angelo, Texas. Jody was there with me one day in late 1980 when one of my critics, Bob Steger (a professor at Angelo State University) knocked on the door. Bob had come, he explained, because three times I had visited their research station, witnessed their grazing and brush clearing experiments, and—without asking for any data—accurately predicted the outcomes. Bob said that each time I left, they had ridiculed me for my lack of interest in their data, yet I hadn't been wrong. Bob's question was simply, "What are you doing that we don't understand?"

In my effort to explain how I could predict their results without needing to see their data, I grabbed pencil and paper, drew four boxes and labeled them. "These," I pointed out, "represent the processes through which our ecosystem functions: the water cycle, mineral cycle, energy flow, and succession (later community dynamics)."[3] Bob nodded and waited.

[3] Most scientific and lay publications refer to many different ecosystems and imply that they function differently. I view us living in one biosphere produced and kept stable by an ecosystem operating through four processes in millions of environments on land and sea.

I continued, "I've learned over the years that we can only manage any environment effectively through these four processes. Realistically, they aren't separate because anything we do to enhance or damage one process affects all of them. It's as if we are looking into the same room through four different windows. There will generally be one window that reveals what most needs attention; once we fix the problem spotted through that window, the other three processes also improve."

I went on to share my many experiences in coming to this realization. While working for the Southern Rhodesian Game Department, I had observed that our management can impair or improve the water that cycles between sky, soil, plants, and oceans, and subsequently I coined the terms *effective* and *noneffective rainfall.* On Kapombe Farm, I had learned that our management could slow or speed up the mineral nutrients cycling between plants, animals, and soil life. The yellowing, mineral-deficient sugarcane I'd planted in mineral-rich basalt soil turned green and grew rapidly when I stopped over-watering, as did the eight-pound-versus-one-pound cabbages I grew to quickly check whether the extension advice I'd been given, on where to plant my sugarcane, was correct.

In working with my ranching clients, I had witnessed how sunlight energy could increasingly be converted to green, growing plants when livestock trampled old forage to make way for fresh, new growth. And from my days in the Northern Rhodesian Game Department, I had come to recognize that the old ecological term *succession*—which inferred a steady progression on land from bare ground through grasslands to forest—was too simple. The progression was anything but steady. Thousands of buffalo could pass through an area and transform bare ground to grassland in a season, or they could be removed from dense grassland, and it would begin to die in a season.

But how could we manage these four processes to produce the desired result? On my sketch, underneath the four boxes, I now listed a handful of "influences" we could manipulate to produce predictable results (favorable or unfavorable) in each of the four processes, emphasizing that a change in one affected them all. I said, "To increase energy flow, for example, we could remove the dead vegetation choking new growth by lighting a *fire* (but this also exposes soil, making the available rainfall less effective); using some form of *technology*, such as mowing (but this is costly and polluting); or using *livestock* to graze and trample the vegetation down and leave manure behind—acting as fertilizer to benefit the new growth."

Now I divided *livestock* into two separate tools: *grazing*—to cycle plant material and keep plants productive—and *animal impact* (the concentrated trampling, dunging, and urinating provided by grazing animals in the presence of pack-hunting predators)—to provide the mulch and seedbed needed for plants to grow and to make the available rainfall more effective. "Here in Texas," I emphasized, "it would be impossible to reverse the desertification without using these last two tools."

Bob was still paying attention, which was promising. Traditionally, American range management professionals (Bob included) considered it heresy to use cattle, sheep, goats and other livestock as a tool for reclamation. Most of them, including those in universities, were adamant that this was wrong—after all, livestock caused desertification! They could not begin to consider livestock as anything more than a destructive force (a view our veld management specialists in Rhodesia had also shared). Bob and his colleagues were still promoting reductions in livestock numbers to eliminate overgrazing and to protect soils from overtrampling. I pointed out to Bob, "Every time you showed me your experiments, I could see the tools your team was using and how you were applying them. From that, I could draw conclusions about the likely influence of each tool on the four ecosystem processes. The likely result differs with the distribution pattern of humidity throughout the year."

Largely gleaned from my days of tracking and observing the land, I had come to understand that the distribution of humidity is more important than the annual rainfall. I could determine the humidity distribution by observing its effects—the proportions of algae, lichen, or moss colonizing the surface of large stones or the bark of trees; how widely plants were spaced; whether the breakdown of old grass was biological (by living organisms) or chemical (through oxidation); and in the degree of soil surface capping or crusting. As soon as they told me what actions (tools) their research was testing, I could predict the results simply by knowing the likely effects those tools would produce in their environment, as defined by how regular or erratic I observed the humidity to be. While this was clear to me from decades of tracking as an ecologist, it was "wizardry" to Bob and his US range science colleagues. To help Bob understand why this guideline and the others were so useful, I drew connecting lines from the guidelines to the tools to the ecosystem processes, showing how the outcome could be predicted with reasonable accuracy.

This all took a couple of hours, and Bob left seemingly more confused than when he arrived. However, Jody, in observing my attempts at sketching out on paper the logic and evidence I used to reach specific conclusions, said, "You need to share what you just did—no one understands how you work these things out."

Share I did; with the help of several thousand individuals from many disciplines and several countries, my sketch that day became the foundation of what we now refer to as the Holistic Management framework, which enables anyone to manage holistically in any situation—as a family, a business, a government, and more. I served as the collator and synthesizer of all this glorious input. I was determined to give the remainder of my life to solving the problems arising from our *lack* of ability to manage holistically, that is, to manage our lives, our food production, and our economies while regenerating our life-supporting environment.

Some months after Bob's visit, Jody and I relocated to New Mexico. Ray Margo and Don Sylvester from the Albuquerque office of the Soil Conservation Service (since renamed the Natural Resources Conservation Service) called on me. They said they had attended a field day at a Texas research station and that the range scientists told them to ignore my work—there was simply no science behind it. At the end of the field day, a rancher had asked, "Why is it that when I do what Savory says, it works, but when you do it on your research stations you say it doesn't work?"

The range scientists didn't have a convincing answer, and Ray and Don, having seen the results themselves on several ranches, were now asking if I would provide training to the Soil Conservation Service.

For me, the response to Ray and Don's request was a no-brainer. Ever since I had first asked Oliver West (at the time the Rhodesian government's chief veld management expert) to take over my work, I had tried to convince others with greater resources to take over its further development due to its global importance. Providing courses for a government agency—and possibly getting the American government to support the work—was beyond my wildest dreams. I readily agreed to teach courses, insisting only that the sessions be open to all agencies and agricultural (land grant) universities.

The Soil Conservation Service's regional office promptly followed up—and contracted me to teach staff from all the US land management agencies: the Soil Conservation Service, US Forest Service, Bureau of Land Management, and Bureau of Indian Affairs. Over the next few years, I facilitated weeklong courses for more than 2,000 individuals from these agencies as well as from the US Fish and Wildlife Service, US Agency for International Development (USAID), and the faculties of United States land grant universities. I even held courses for representatives from the World Bank, who subsequently engaged me as a consultant.

These became exciting years of incredible growth and discovery. The courses, each with about thirty participants, followed a similar progression. The rough management framework that I had first sketched for Bob became the focal point. I continually made improvements to it as I learned simpler and easier ways for everyone to understand. As atonement for the elephant culling catastrophe of the 1960s, I did everything possible to ensure nothing I taught—or developed together with the participants—went unchallenged. The first hour of each day's session was devoted exclusively to challenging both the science and the logic in everything we did to further develop this management framework. My sketch grew rapidly into something much more robust, user friendly, and easily taught. I added a key element—which later became known as a *holistic context*—to drive management and policy making. The holistic context tied our lives and livelihoods to our life-sustaining environment and described the situation beyond immediate concern. Now—as we met our needs or solved our problems—we did so within the holistic context, never losing sight of what was ultimately most important to us.

As part of the courses, I insisted that participants share their own resource management policies for us to analyze, just as I had done with Bob Steger's research projects.

These policies covered every resource management problem imaginable, from droughts and floods, invasive and poisonous plants, and insect and rodent infestations to silting dams, biodiversity loss, the death of small towns, and more. The agency and university participants—upon analyzing their own policies using the new framework—had difficulty finding any that would not fail or that would not lead to unintended adverse consequences. One group made a joint statement we recorded and later published, "We now recognize that unsound resource management is universal in the United States."[4] These failures, we discovered, were not for lack of scientific knowledge or good intent, but due to the way the policies were being developed.

We began to use the Holistic Management framework to revise policies. In many cases we found that with minimal change we could make the policy far more likely to succeed. This was eye-opening to everyone who participated. Never had I thrived more intellectually than in working with so many bright, caring minds across agencies, disciplines, and universities.

Don chaired an Interagency Committee (IAC) to take on the administrative side of the courses so I could concentrate on the teaching. My agreement with the IAC was that they would monitor the response. If feedback was favorable—and the demand for teaching amongst the agencies grew—the US government would then take over, creating a training center to carry on my work. I could then return to Zimbabwe, as I wished to do.

Not every session went smoothly, especially among those participants who took criticism of their policies personally. Those few—often eminent academics—reacted with anger, sometimes extreme anger. While none of them pointed out flaws in the logic or science, they did point out plenty of flaws in me—my offensive manner, my Rhodesian accent, the way I dressed, my motive, and so forth. However, it was new knowledge that was the threat, not me. The experience was turning their worlds upside down and heading in the opposite direction of the one they'd published a career's worth of papers on. They did not share the expectation, drummed into us at the University of Natal, that judging scientific work was an essential part of the game—not to be confused with judging scientists as human beings.

The IAC and I were about to learn a lesson we should have been aware of but would in fact not understand for many years. There are essentially two types of discoveries. The most common are those that are seen by everyone as "new" discoveries—no one has previously known about or understood them. Examples include advances in areas of life that were regarded important enough to warrant Nobel Prizes and subsequently—after numerous discussions, differences, arguments, experimentations, and replications—received institutional acceptance.

[4] This statement has been published in all three editions of my textbook, *Holistic Management* (Island Press: Washington D.C., 1988, 1999, 2016).

The other type of discovery, which is rare and far less obvious, is one that some people may see as new, but that experts and institutions, which reflect the beliefs of the society in which they were formed, know with certainty is not new but *wrong*. This type of discovery goes against centuries, or even millennia, of unquestioned belief. Examples include the discoveries of Copernicus and Galileo[5]—their thinking was declared "wrong" by church and state leaders who *knew* the world was flat and that the Sun revolved around the Earth, rising at dawn and disappearing before dusk.

It is institutions—whether government, academia, environmental organizations, churches, or even a corporation—that promote the ridicule and rejection of the new "heresy." And sadly, they will do so for a very long time, no matter how much data is provided, how many lives are lost, nor how high the environmental or economic cost. Only when the beliefs of society change significantly—up until this point a period measured in centuries—do institutions come around to accepting the new discovery.[6]

Not being aware of the downfalls of this second type of discovery, the IAC and I were completely caught off guard when things suddenly fell apart. While most participants had enthusiastically embraced the work and we were in the process of planning to teach thousands more (including 18,000 from the US Forest Service alone), those participants who had reacted with extreme anger were at the pinnacle of their professions and were not about to accept this "new" work that they "*knew* was wrong" as valid. They had an immense amount of influence in the Reagan administration and, to our dismay, insisted there would be no government training center nor funding available to expand the training as we had been planning.

Don called an IAC meeting and apologized for how powerless they were to do anything to sway the higher echelons of academia and government. As Don explained, my agreement with the government could no longer be honored. Not only would there not be a government training center, but we were likely to be shut down altogether. That said, the committee believed it vital for the future of the United States that the work continue. To ensure this, they urged me to form a nonprofit organization to continue the training and to further the development of Holistic Management. A nonprofit organization, they said, could not be shut down, even by the President. However, I would have to form it (as government agency staff, they could not).

Not being American—and wanting to return home—this was a devastating appeal. I asked for a day to consider it. Jody and I talked it over and concluded that I had to step up. It was this that ultimately led to the formation of what is today the

---

[5] Nicolaus Copernicus and Galileo Galilei were pivotal figures in the Copernican Revolution, a shift in astronomical thought that placed the Sun—not Earth—at the center of our solar system. Copernicus proposed the heliocentric model, while Galileo provided observational evidence—particularly through his use of telescopes—to support this theory.

[6] For more on the topic of discoveries, see Eric Ashby's *Reconciling Man with the Environment* (London: Oxford University Press, 1978) and Everett Rogers's *Diffusion of Innovations* (New York: Simon and Schuster, 2003).

Savory Institute, where the global development of Holistic Management continues to be coordinated.[7]

Two years later, in 1986, after we had formed the initial nonprofit training organization (now Savory Institute), the IAC was dissolved by the second Reagan administration's newly appointed Soil Conservation Service Chief, Wilson Scaling. Staff from all the government land management agencies were prohibited from taking further training.

Shortly following this, I had a conversation with Professor Jim Teer of Texas A&M University, who had followed the demise of the IAC and our courses. He said to me, "Allan, either you are wrong, and the world won't be able to dig a hole deep enough to bury you in, or you are right, and the world won't be able to build a monument high enough."

I quickly corrected Jim, emphasizing that this work wasn't about me. "What I want to know," I said, "is what do *you* think?" Jim said he was "sitting on the fence," unable to make up his mind on whether I was right or wrong. Jim died sitting on that fence. Many more have sat on the fence—for nearly half a century, millions of people have suffered through floods and droughts, hunger, poverty, and violence, as the reality of biodiversity loss, desertification, and accelerating climate change continues.

In 1975 my family's home, Kazuma, became part of Kazuma Pan National Park. I hold many a happy memory from my days there, especially the times I spent with Jim and Meg. It was a dream come true to have been able to hunt with Jim and ride horseback with Meg across the swaths of African savanna teeming with the wildlife I so love.

Meg went on to become a successful veterinarian and competitive endurance rider in the United States. Jim, after being taken to South Africa by his mother, joined the South African Army. He was commissioned and served as a company commander in the renowned 32 "Buffalo" Battalion operating in Angola, where he was twice wounded. I persuaded him to leave the army (as he did not believe in what the apartheid government was trying to do) and come to the United States to start life anew. Several months after starting work on a ranch in New Mexico, Jim was killed in an accident. Just days prior, he had joined Jody and me on a weekend break. We spent much of that time telling stories and laughing; I loved seeing him so engaged and happy. My two most vivid memories of Jim are of that day, and of the first time I held him in my arms.

Claire, though she lived most of her life with her mother, came to live for a few years with Jody and me, by then a troubled albeit loving child who gave us joy amidst the challenges. She left this world far too early. Her brother, Rodger, was sent to live with us at age sixteen, arriving confused and angry. Even with help I was unable to develop the closeness with him I hoped for, and to this day I struggle to understand him. Yet, we have enjoyed some good times together.

[7] For more information on the Savory Institute, visit https://savory.global.

Jim and Claire are both buried at Kazuma. In 2021, Meg brought Shirley's ashes to Kazuma as well, spreading them around the site of our original campfire, where we slept on the ground in the early days before we had built our mud and thatch huts.

I got to see my youngest daughter, Sarah, on my first trip back to Zimbabwe. She was nearly seven years old by then and had little memory of me or of how close we had been four years earlier. Every few years thereafter, when I went to Zimbabwe, I would visit, albeit fleetingly. I appeared to her a rather boring, older person and she quickly rushed off to play with her (far more interesting) friends. I got the chance to know Sarah somewhat much later in her life, when she had two young children she felt might like to know their grandfather—and for that I've been grateful.

From a family perspective, my largest regret is that I failed in my prime to create for my children—and those of all Rhodesians, Black and White—a prosperous, healthy Zimbabwe in which all people could feel secure and well governed.

Shortly after returning to Zimbabwe from exile, I made the decision to sell Gumela, my beautiful ranch in the Matopos Hills where I'd hoped to retire. I could not justify keeping it for that purpose, so I sold it to the government for settlement. I still had one remaining ranch, Waterford (the 6,000-acre spread which included an additional 13,000 contiguous acres known as Waterford Extension that I had leased from the government with an option to purchase). In the early 1990s Jody and I, with agreement from Prime Minister Mugabe, donated Waterford and its extension to the Dimbangombe Trust.[8] We had formed this trust with the five local chiefs as permanent trustees. Our aim was that it be used as a place where people from all over Africa could come to learn about Holistic Management and experience it in practice. Approximately 2,000 acres of Waterford Extension was taken by settlers during the invasions of "stolen land" in the late 1990s, and another 6,000 acres taken by the governor of Matabeleland and given to two of her political supporters (who lease it to hunters). Yet Dimbangombe, as we now refer to Waterford, continues to serve as a base from which Jody and I live and work to the backdrop of my beloved Zimbabwean wildlife and people.

As I close this chapter of my life, I am immensely grateful for the many opportunities I've had, and continue to have, working with thousands of ranchers, farmers, pastoralists, fellow scientists, and academics to first understand and then resolve the twin problems of biodiversity loss and desertification, which are fueling climate change and threatening all of humanity. My belief over sixty years ago—that our management as government agents and experts was somehow causing the environmental degradation I was seeing—led to years of rejection. Now, thankfully, almost all scientists agree that humans are directly responsible. Logically, this means that human management is at fault, not the things we

[8] The Dimbangombe Trust was originally known as the Africa Centre for Holistic Resource Management Trust.

usually blame. It is how we *manage* livestock, not the animals themselves that is at fault, and it is how we *manage* the extraction and use of coal, gas, and oil resources that is at fault, not the resources themselves. The tie to human management has, as I write, still not penetrated the media, the majority of the scientific community, nor society at large. Thus, world leaders continue to believe that fossil fuels and livestock cause desertification and climate change, not our management of them, and their policies reflect it. That leaves future generations facing disasters we cannot begin to imagine, if we don't change this very soon.

While my efforts to resolve the greatest dangers I could see in my country resulted in astounding opposition from people in way more powerful and influential positions than I, this only strengthened my resolve. This is not because I am bullheaded or arrogant but because I know what I know, and that includes my place in the world.

In the greater scheme of things, I am meaningless—every atom in me has cycled for millennia through soils, plants, insects, birds, reptiles and more, and will return to doing so in a very short time. It is only my mind that is me alone and that, too, will last only briefly. Hence, I've continually tried to challenge and maximize the use of my mind. No matter how much I might be ridiculed or hated, even if it were coming from an entire country that regarded me as unsavory, I couldn't live with myself if I ever did what I knew to be wrong for all of us.

While still a schoolboy at Plumtree, thinking about my future, I knew I wanted to live an active and meaningful life, and subsequently chose professions that provided it, though fraught with conflict and setbacks. But even so, I would come to liken my life to a game of cricket. I would only have one innings, one chance to bat for my team—team humanity. At some point I would be declared out by the umpire—caught, stumped, bowled, leg before wicket, or run out—so I had better do the best I could for my team, not knowing how long my innings would be. And that I have strived to do.

ACKNOWLEDGMENTS

I am first and foremost indebted to the generations of hunter gatherers who came before us. Their skill in tracking animals, making observations, and reasoning about the natural world led to the domestication of plants and animals—the foundation that made civilization and modern science possible.

I would also like to acknowledge the 18th century naturalist, explorer, and polymath Alexander von Humbolt, who was recognized posthumously as the father of ecology—the branch of science on which the future of all economies, businesses, and cities depends. Alongside von Humbolt I acknowledge Jan Smuts, who gave us the concept of holism—the theoretical base of my work. And I am indebted to my friends and mentors mentioned in this memoir: Eustace Poles, Blake Goldsmith, Frank Ansell, Sir Frank Fraser Darling, Archie Mossman, and Sir Roy Welensky. They made me a better scientist and critical thinker.

Eight years ago, my wife, Jody—out of concern that I would never write this memoir (and my stories would then be forgotten)—convinced our friends Gretel Ehrlich and her husband, the late Neal Conan, to join us around our campfire in Zimbabwe. Together we recorded nearly thirty hours of my storytelling. This gave us the rough material that eventually became a first draft of this book.

My long-time friend and collaborator Sam Bingham read through early drafts. With his formidable skills as both writer and editor—and his belief in the book's value—he helped create a flow to the narrative, prodding me throughout to explain myself. In the process he reacquainted me with the thinking of Carl von Clausewitz. To Sam, we owe the version of Clausewitz' Trinity that appears in this book. Sam translated Clausewitz's words from the original German—words which influenced my own understanding of why citizens, combatants, and politicians support war.

Editor Jamie Cross helped make the book more reader friendly by pointing out that some of the characters and ideas mentioned had long faded from public memory and were incomprehensible to people who didn't live through those times. The restructuring of lengthy chapters, the sprinkling of footnotes throughout, and the addition of maps, timelines, and photos were her suggestions. As it turns out, they were much needed.

Because I spent much of my life in motion, I had limited opportunities for keeping a diary or taking photos. The photos in this book are mostly drawn from my archives. If I didn't take them myself, I've done my best to credit the photographer. In some cases, the photos and/or formerly published pieces are from so long ago that I've been unable to identify the source nor solidify permission to use them, despite exhaustive efforts and research.

In the earliest days, on a 300-mile walk, for documentation purposes I could carry no more than a small diary and two roles of black and white film (which limited me to sixteen photos). I had to rely on my memory for an immense number of details. I am indebted to three historians of Southern Africa: Shaun Milton, Brooks Marmon, and Kevin Thomas. They reviewed some or all of the early chapters, rectified embarrassing errors, and helped reorder events as they "actually" occurred. I am accountable for any errors that remain.

Authors Peter Godwin, John Padbury, the late Jim Latham, Gretel Ehrlich, Judy Schwartz, and Tony Eprile have written about some of the same events I describe. In turn they reviewed various chapters and early drafts of this book, making suggestions that improved it immensely. Many friends who lived through some of the events—including Archie Mossman, Colleen Fairlamb Duffy, Brian Robinson, Alun Vaughan, and my daughter Megan Savory-Davis—also read through early drafts or specific chapters, matching their recollections with my own, and providing candid feedback when they differed. I am also grateful to the late Ken Flower and the late Ian Smith for writing memoirs before I did; they corroborate some of the stories I tell that would otherwise have been unbelievable.

Finally, my love and thanks to my wife, partner, and collaborator, Jody Butterfield, to whom I've also dedicated this book. Without her encouragement, determination, skills, and patience this memoir would never have come into existence.

INDEX

NOTE: Page numbers in *italics* indicate an image, and an *n* after a page number indicates a footnote.

E

F

O

P

Q

R

S

T

9 798993 545400

Printed by Libri Plureos GmbH in Hamburg,
Germany